Pre-Algebra
DeMYSTiFieD®

DeMYSTiFieD® Series

Accounting Demystified
Advanced Calculus Demystified
Advanced Physics Demystified
Advanced Statistics Demystified
Algebra Demystified
Alternative Energy Demystified
Anatomy Demystified
asp.net 2.0 Demystified
Astronomy Demystified
Audio Demystified
Biology Demystified
Biotechnology Demystified
Business Calculus Demystified
Business Math Demystified
Business Statistics Demystified
C++ Demystified
Calculus Demystified
Chemistry Demystified
Circuit Analysis Demystified
College Algebra Demystified
Corporate Finance Demystified
Data Structures Demystified
Databases Demystified
Differential Equations Demystified
Digital Electronics Demystified
Earth Science Demystified
Electricity Demystified
Electronics Demystified
Engineering Statistics Demystified
Environmental Science Demystified
Everyday Math Demystified
Fertility Demystified
Financial Planning Demystified
Forensics Demystified
French Demystified
Genetics Demystified
Geometry Demystified
German Demystified
Home Networking Demystified
Investing Demystified
Italian Demystified
Java Demystified
JavaScript Demystified
Lean Six Sigma Demystified
Linear Algebra Demystified

Logic Demystified
Macroeconomics Demystified
Management Accounting Demystified
Math Proofs Demystified
Math Word Problems Demystified
MATLAB® Demystified
Medical Billing and Coding Demystified
Medical Terminology Demystified
Meteorology Demystified
Microbiology Demystified
Microeconomics Demystified
Nanotechnology Demystified
Nurse Management Demystified
OOP Demystified
Options Demystified
Organic Chemistry Demystified
Personal Computing Demystified
Pharmacology Demystified
Physics Demystified
Physiology Demystified
Pre-Algebra Demystified
Precalculus Demystified
Probability Demystified
Project Management Demystified
Psychology Demystified
Quality Management Demystified
Quantum Mechanics Demystified
Real Estate Math Demystified
Relativity Demystified
Robotics Demystified
Sales Management Demystified
Signals and Systems Demystified
Six Sigma Demystified
Spanish Demystified
SQL Demystified
Statics and Dynamics Demystified
Statistics Demystified
Technical Analysis Demystified
Technical Math Demystified
Trigonometry Demystified
UML Demystified
Visual Basic 2005 Demystified
Visual C# 2005 Demystified
XML Demystified

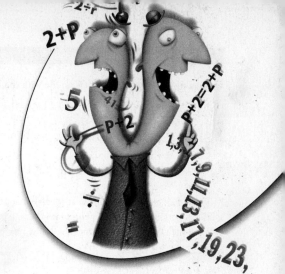

Pre-Algebra
DeMYSTiFieD®

Allan G. Bluman

Second Edition

McGraw Hill

New York   Chicago   San Francisco   Lisbon   London   Madrid   Mexico City
Milan   New Delhi   San Juan   Seoul   Singapore   Sydney   Toronto

The **McGraw·Hill** Companies

Cataloging-in-Publication Data is on file with the Library of Congress

Pre-Algebra DeMYSTiFieD®, Second Edition

2 3 4 5 6 7 8 9 0   DOC/DOC   1 6 5 4 3 2 1

ISBN 978-0-07-174252-8
MHID 0-07-174252-2

Sponsoring Editor
Judy Bass

Editorial Supervisor
Stephen M. Smith

Production Supervisor
Pamela A. Pelton

Acquisitions Coordinator
Michael Mulcahy

Project Manager
Anupriya Tyagi,
Glyph International

Copy Editor
Namita Panda,
Glyph International

Proofreader
Manish Tiwari,
Glyph International

Cover Illustration
Lance Lekander

Art Director, Cover
Jeff Weeks

Composition
Glyph International

To Brooke Leigh Bluman

About the Author

Allan G. Bluman taught mathematics and statistics in high school, college, and graduate school for 39 years. He received his doctor's degree from the University of Pittsburgh. He has written three mathematics textbooks published by McGraw-Hill. In addition, he has written three other *Demystified* books in mathematics. Dr. Bluman is the recipient of an "Apple for the Teacher" award for bringing excellence to the learning environment and the "Most Successful Revision of a Textbook" award from McGraw-Hill. His biographical record appears in *Who's Who in American Education*, Fifth Edition.

Contents

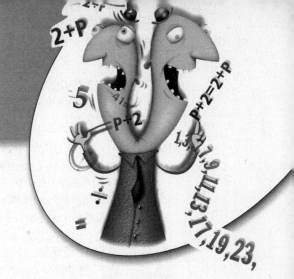

Preface

In order to be successful in mathematics, it is necessary to have a strong foundation. This foundation consists of the mastery of the basic concepts of arithmetic, which includes whole numbers (Chap. 1), fractions (Chaps. 3 and 4), decimals (Chap. 5), and percents (Chap. 6). These topics are the foundation of this book.

Since most students go on to study algebra, the basic concepts of algebra, consisting of the integers (Chap. 2), evaluation of expressions and equations (Chap. 7), graphing (Chap. 11), and operations with monomials and polynomials (Chap. 12), are also included here. Finally, the topics of ratio and proportion (Chap. 8), informal geometry (Chap. 9), and measurement (Chap. 10) have been included in order to help students with everyday applications of problem solving.

A tutorial on study skills and overcoming math anxiety is presented in the Appendix. Part I explains the nature and causes of math anxiety. Part II contains suggestions for overcoming math anxiety. Part III lists study and test-taking tips for success in mathematics. Even if you don't have math anxiety, you should read and use the suggestions given in Part III. It is written for all students. You must realize that mathematics requires analytical reasoning, problem-solving skills, and critical thinking. There is much more to understanding mathematics than just memorization.

Curriculum Guide

The *Demystified*® books are closely linked to the standard high school and college curricula, so the Curriculum Guide found on the inside back cover is provided to give you a clear path to meet your mathematical goals. What many students do not know is that mathematics is a hierarchical subject. What this means is that before you can be successful in algebra, you need to know basic arithmetic, since the concepts of arithmetic (pre-algebra) are used in algebra. Before you can be successful in trigonometry, you need to have a basic understanding of algebra and geometry, since trigonometry uses concepts from these two courses. You can use the Guide in your mathematical studies to learn which courses are necessary before you take the next one.

How to Use This Book

As you know, in order to build a tall building, you need to start with a strong foundation. It is also true in mastering mathematics that you need to start with a strong foundation. This book presents the basic topics in arithmetic and introductory algebra in a logical, easy-to-read format. This book can be used as an independent study course or as a supplement to a pre-algebra course.

To learn mathematics, you must know the vocabulary, understand the rules and procedures, and be able to apply these rules and procedures to mathematical problems in order to solve them. This book is written in a style that will help you with learning. Important terms have been boldfaced and important rules and procedures have been italicized. Basic facts and helpful suggestions can be found in Math Notes. Each section has several worked-out examples showing you how to use the rules and procedures. Each section also contains several practice problems for you to work out to see if you understand the concepts. The correct answers are provided immediately after the problems so you can see if you have solved them correctly. At the end of each chapter is a 20-question multiple-choice quiz. If you answer most of the problems correctly, you can move on to the next chapter. If not, please repeat the chapter. Make sure you do not look at the answer before you have attempted to solve the problem.

Even if you know some or all of the material in a chapter, it is best to work through the chapter in order to review the material. The little extra effort will be a great help when you encounter more difficult material later. After you complete the entire book, you can take the 100-question final exam and determine your level of competence.

It is suggested that you do *not* use a calculator since a calculator is only a tool, and there is a tendency to think that if a person can press the right buttons and get the correct answer, then the person *understands* the concepts. This is far from the truth!

I would like to answer the age-old question, "Why do I have to learn this stuff?" There are several reasons. First, mathematics is used in many academic fields. If you cannot do mathematics, you severely limit your choices of an academic major. Second, you may be required to take a standardized test for a job, degree, or graduate school. Most of these tests have a mathematics section. Third, a working knowledge of arithmetic will go a long way to help you to solve mathematical problems that you encounter in everyday life. I hope this book will help you to learn mathematics.

For this second edition, most of the examples and exercises have been changed. In addition, I have added over 20 new Math Notes to make the material easier to follow, and I have added a section in Chap. 3 on divisibility rules to help with the topic of reducing fractions. Finally, an explanation of two additional grouping symbols, brackets and braces, has been included in Chap. 7.

Best wishes on your success!

Allan G. Bluman

Acknowledgments

I would like to thank my wife, Betty Claire, for helping me with this project, and I wish to express my gratitude to my editor, Judy Bass, and to Carrie Green for their assistance in the publication of this book.

chapter 1

Whole Numbers

Numbers make up the foundation of mathematics. The first numbers people used were the natural or counting numbers, consisting of 1, 2, 3, ... When 0 is added to the set of natural numbers, the set is called the whole numbers. This chapter explains the basic operations of addition, subtraction, multiplication, and division of these numbers.

CHAPTER OBJECTIVES

In this chapter, you will learn how to

- Read whole numbers
- Round whole numbers
- Add, subtract, multiply, and divide whole numbers
- Solve word problems using whole numbers

Naming Numbers

Our number system is called the **Hindu-Arabic** system or **decimal** system. It consists of 10 symbols or **digits**, 0, 1, 2, 3, 4, 5, 6, 7, 8, and 9, which are used to make our numbers. Each digit in a number has a **place value**. The place value names are shown in Fig. 1-1.

Place Values														
Trillions			Billions			Millions			Thousands			Ones		
Hundred trillions	Ten trillions	Trillions	Hundred billions	Ten billions	Billions	Hundred millions	Ten millions	Millions	Hundred thousands	Ten thousands	Thousands	Hundreds	Tens	Ones

FIGURE 1-1

In larger numbers each group of three numbers (called a **period**) is separated by a comma. The names at the top of the columns in Fig. 1-1 are called period names.

To name a number, start at the left and go to the right, read each group of three numbers separately using the period name at the comma. The word "ones" is not used when naming numbers.

EXAMPLE

Name 62,432,709.

SOLUTION

Sixty-two million, four hundred thirty-two thousand, seven hundred nine.

EXAMPLE

Name 560,711.

SOLUTION

Five hundred sixty thousand, seven hundred eleven.

EXAMPLE

Name 87,001,000,012.

SOLUTION

Eighty-seven billion, one million, twelve.

MATH NOTE *Other period names after trillions in increasing order are quadrillion, quintillion, sextillion, septillion, octillion, nonillion, decillion, and so fourth.*

Practice

Name each number:

1. 515
2. 27,932
3. 1,607,003
4. 63,902,400,531

Answers

1. Five hundred fifteen
2. Twenty-seven thousand, nine hundred thirty-two
3. One million, six hundred seven thousand, three
4. Sixty-three billion, nine hundred two million, four hundred thousand, five hundred thirty-one

Rounding Numbers

Many times it is not necessary to use an exact number. In this case, an approximate number can be used. Approximations can be obtained by rounding numbers. All numbers can be rounded to specific place values.

To round a number to a specific place value, first locate that place value digit in the number. If the digit to the right of the specific place value digit is 0, 1, 2, 3, or 4, the place value digit remains the same. If the digit to the right of the specific place value digit is 5, 6, 7, 8, or 9, add one to the specific place value digit. In either case all digits to the right of the specific place value digit are changed to zeros.

EXAMPLE

Round 52,183 to the nearest hundred.

SOLUTION

We are rounding to the hundreds place which is the digit 1. Since the digit to the right of the 1 is 8, add 1 to 1 to get 2. Change all digits to the right to zeros. Hence 52,183 rounded to the nearest hundred is 52,200.

EXAMPLE

Round 53,462 to the nearest thousand.

SOLUTION

We are rounding to the thousands place, which is the digit 3. Since the digit to the right of the 3 is 4, the 3 stays the same. Change all digits to the right of 3 to zeros. Hence 53,462 rounded to the nearest thousand is 53,000.

EXAMPLE

Round 1,498,352 to the nearest ten thousand.

SOLUTION

We are rounding to the ten thousands place, which is the digit 9. Since the digit to the right of the 9 is 8, the 9 becomes a 10. Then we write the 0 and add the 1 to the next digit to the left. The 4 then becomes a 5. Hence the answer is 1,500,000.

Still Struggling

Remember to change digits to the right of the rounding place to zero when rounding. For example, when you round 4,672 to the nearest hundred, you get 4,700 not 47.

Practice

1. Round 7,831 to the nearest thousand.
2. Round 294,183 to the nearest ten thousand.
3. Round 92,308 to the nearest ten.
4. Round 682,611 to the nearest hundred thousand.
5. Round 163,793,244 to the nearest million.

Answers

1. 8,000
2. 290,000
3. 92,310
4. 700,000
5. 164,000,000

Addition of Whole Numbers

In mathematics, addition, subtraction, multiplication, and division are called **operations**. The numbers being added are called **addends**. The total is called the **sum**.

$$
\begin{array}{r}
5 \leftarrow \text{addend} \\
+ 2 \leftarrow \text{addend} \\
\hline
7 \leftarrow \text{sum}
\end{array}
$$

To add two or more numbers, first write them in a column, and then add the digits in the columns from right to left. If the sum of the digits in any column is 10 or more, write the one's digit and carry the ten's digit to the next column to the left and add it to the numbers in that column.

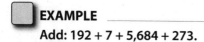**EXAMPLE**

Add: 192 + 7 + 5,684 + 273.

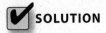 **SOLUTION**

```
         121   ← carry row
         192
           7
        5684
   +     273
        6156
```

MATH NOTE *To check addition, add from the bottom to the top.*

```
         273
        5684
           7
   +     192
        6156
```

Practice
Add:

1. 443 + 27 + 7
2. 4,593 + 14 + 863
3. 7,324 + 625,713
4. 18 + 46,933 + 36 + 557
5. 5,641 + 300 + 65 + 77,325

Answers
1. 477
2. 5,470
3. 633,037
4. 47,544
5. 83,331

Subtraction of Whole Numbers

In subtraction, the top number is called the **minuend**. The number being sub-tracted (below the top number) is called the **subtrahend**. The answer in sub-traction is called the **remainder** or **difference**.

$$
\begin{array}{r}
875 \\
-\ \ 43 \\
\hline
832
\end{array}
\quad
\begin{array}{l}
\leftarrow \text{ minuend} \\
\leftarrow \text{ subtrahend} \\
\leftarrow \text{ difference}
\end{array}
$$

To subtract two numbers, write the numbers in a vertical column and then sub-tract the bottom digits from the top digits. Proceed from right to left. When the bottom digit is larger than the top digit, borrow one from the digit at the top of the next column and add ten to the top digit before subtracting. When borrowing, be sure to reduce the top digit in the next column by 1.

 EXAMPLE

Subtract: 19,784 – 4,213.

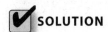 **SOLUTION**

$$
\begin{array}{r}
19784 \\
-\ \ \ 4213 \\
\hline
15571
\end{array}
$$

 EXAMPLE

Subtract: 5,386 – 748.

 **SOLUTION**

$$
\begin{array}{r}
4\ \ 13\ \ \ 7\ \ 16 \quad \leftarrow \textbf{ borrowing numbers} \\
\cancel{5}\ \ \cancel{3}\ \ \cancel{8}\ \ \cancel{6} \\
7\ \ 4\ \ 8 \\
\hline
4\ \ 6\ \ 3\ \ 8
\end{array}
$$

MATH NOTE   *To check subtraction, add the difference to the subtrahend to see if you get the minuend.*

$$
\begin{array}{r}
5386 \\
-\ \ 748 \\
\hline
4638
\end{array}
\qquad
\begin{array}{r}
\text{Check} \\
4638 \\
+\ \ 748 \\
\hline
5386
\end{array}
$$

Practice

Subtract:

1. 961 – 87
2. 24,271 – 6,314
3. 867,281 – 23,779
4. 73,307,641 – 863,259
5. 8,000,000 – 81,406

Answers

1. 874
2. 17,957
3. 843,502
4. 72,444,382
5. 7,918,594

Multiplication of Whole Numbers

In multiplication, the top number is called the **multiplicand**. The number directly below it is called the **multiplier**. The answer in multiplication is called the **product**. The numbers between the multiplier and the product are called **partial products**.

$$
\begin{array}{r}
425 \\
\times\ \ \ \ 63 \\
\hline
1275 \\
2550\ \ \\
\hline
26775
\end{array}
\quad
\begin{array}{l}
\leftarrow \text{multiplicand} \\
\leftarrow \text{multiplier} \\
\leftarrow \text{partial product} \\
\leftarrow \text{partial product} \\
\leftarrow \text{product}
\end{array}
$$

To multiply two numbers when the multiplier is a single digit, write the numbers in a vertical column and then multiply each digit in the multiplicand from right to left by the multiplier. If any of these products is over nine, add the tens digit to the product of numbers in the next column.

EXAMPLE

Multiply: 416 × 7.

SOLUTION

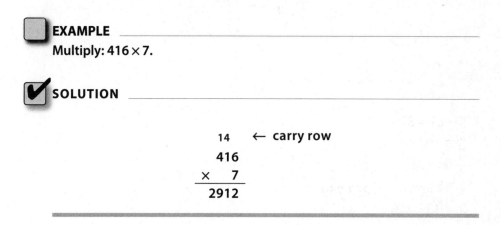

$$
\begin{array}{r}
14 \quad \leftarrow \textbf{ carry row} \\
416 \\
\times \quad 7 \\
\hline
2912
\end{array}
$$

To multiply two numbers when the multiplier contains two or more digits, arrange the numbers vertically and multiply each digit in the multiplicand by the right-most digit in the multiplier. Next multiply each digit in the multiplicand by the next digit in the multiplier and place the second partial product under the first partial product, moving one space to the left. Continue the process for each digit in the multiplier and then add the partial products to get the final product.

EXAMPLE

Multiply: 3,742 × 814.

SOLUTION

$$
\begin{array}{r}
3742 \\
\times \quad 814 \\
\hline
14968 \\
3742 \\
29936 \\
\hline
3045988
\end{array}
$$

MATH NOTE *To check the multiplication problem, multiply the multiplier by the multiplicand.*

$$
\begin{array}{r}
814 \\
\times \quad 3742 \\
\hline
1628 \\
3256 \\
5698 \\
2442 \\
\hline
3045988
\end{array}
$$

Practice

Multiply:

1. 92×5
2. 651×87
3. $4{,}135 \times 216$
4. $61{,}405 \times 892$
5. $154{,}371 \times 43$

Answers

1. 460
2. 56,637
3. 893,160
4. 54,773,260
5. 6,637,953

Division of Whole Numbers

In division, the number under the division box is called the **dividend**. The number outside the division box is called the **divisor**. The answer in division is called the **quotient**. Sometimes the answer does not come out *even*; hence, there will be a **remainder**.

$$
\begin{array}{r}
3 \quad \leftarrow \text{quotient} \\
\text{divisor} \rightarrow \quad 9\overline{)32} \quad \leftarrow \text{dividend} \\
\underline{27} \quad \\
5 \quad \leftarrow \text{remainder}
\end{array}
$$

The process of long division *consists of a series of steps. They are divide, multiply, subtract, and bring down. When dividing, it is also necessary to estimate how many times the divisor divides into the dividend. When the divisor consists of two or more digits, the estimation can be accomplished by dividing the first digit of the divisor into the first one or two digits of the dividend. The process is shown next.*

 EXAMPLE

Divide: 863 by 52.

 SOLUTION

Step 1:

$$\begin{array}{r} 1 \\ 52\overline{)863} \end{array}$$ Divide 5 into 8 to get 1.

Step 2:

$$\begin{array}{r} 1 \\ 52\overline{)863} \\ \underline{52} \end{array}$$ Multiply 1 × 52.

Step 3:

$$\begin{array}{r} 1 \\ 52\overline{)863} \\ \underline{52} \\ 34 \end{array}$$ Subtract 52 from 86.

Step 4:

$$\begin{array}{r} 1 \\ 52\overline{)863} \\ \underline{52} \\ 343 \end{array}$$ Bring down the next digit, 3.

Repeat Step 1:

$$\begin{array}{r} 16 \\ 52\overline{)863} \\ \underline{52} \\ 343 \end{array}$$ Divide 5 into 34.
    (Note: If the answer 6 is too large, try 5.)

Repeat Step 2:

$$\begin{array}{r} 16 \\ 52\overline{)863} \\ \underline{52} \\ 343 \\ \underline{312} \end{array}$$ Multiply 6 × 52.

Repeat Step 3:

$$\begin{array}{r} 16 \\ 52\overline{)863} \\ \underline{52} \\ 343 \\ \underline{312} \\ \overline{31} \end{array}$$ Subtract 312 from 343.

Hence, the answer is 16 remainder 31 or 16 R31. Stop when you run out of digits in the dividend to bring down.

 EXAMPLE

Divide: 4,378 by 67.

 SOLUTION

$$\begin{array}{r} 65 \\ 67\overline{)4378} \\ \underline{402} \\ 358 \\ \underline{335} \\ 23 \end{array}$$

Division can be checked by multiplying the quotient by the divisor, adding the remainder, and seeing if you get the dividend. For the previous example, multiply 65 × 67 and add 23.

$$\begin{array}{r} 65 \\ \times\ \ 67 \\ \hline 455 \\ 390 \\ \hline 4355 \\ +\ \ \ 23 \\ \hline 4378 \end{array}$$

Practice

Divide:

1. $2,494 \div 43$

2. $43,967 \div 7$

3. $40,898 \div 143$

4. $688 \div 31$

5. $10,568 \div 738$

Answers

1. 58

2. 6,281

3. 286

4. 22 R6

5. 14 R236

Word Problems

In order to solve word problems follow these steps:

1. *Read the problem carefully.*

2. *Identify what you are being asked to find.*

3. *Perform the correct operation or operations.*

4. *Check your answer or at least see if it is reasonable.*

In order to know what operation to perform, it is necessary to understand the basic concept of each operation.

MATH NOTE *Always read the word problem at least twice before starting it.*

Addition

When you are asked to find the "sum" or the "total" or "how many in all," and the items are the same in the problem, you add.

EXAMPLE

Find the total calories in a breakfast consisting of a sausage and egg sandwich (450 calories), hash brown potatoes (140 calories), and low-fat milk (95 calories).

SOLUTION

Since you want to find the total number of items and the items are the same (calories), you add: 450 + 140 + 95 = 685 calories. Hence, the total number of calories in the meal is 685.

Subtraction

When you are asked to find the difference, that is, "how much more," "how much less," "how much larger," "how much smaller," etc., and the items are the same, you subtract.

 EXAMPLE

The height of the Matterhorn is 14,690 feet, and the height of Mt. McKinley is 20,320 feet. How much higher is Mt. McKinley than the Matterhorn?

 SOLUTION

Since you are asked how much higher Mt. McKinley is than the Matterhorn, you subtract: 20,320 − 14,690 = 5,630. Hence, Mt. McKinley is 5,630 feet higher than the Matterhorn.

Multiplication

When you are asked to find a total and the items are different, you multiply.

 EXAMPLE

An auditorium consists of 18 rows with 24 seats in each row. How many people can the auditorium seat?

 SOLUTION

Since you want a total and the items are different (rows and seats), you multiply:

$$
\begin{array}{r}
24 \\
\times\ 18 \\
\hline
192 \\
24 \\
\hline
432
\end{array}
$$

The auditorium will seat 432 people.

Division

When you are given a total and are asked to find how many items are in each part, you divide.

 EXAMPLE

If 72 cameras are packed in 6 boxes, how many cameras would be placed in each box?

 SOLUTION

In this case, the total is 72, and they are to be put into 6 boxes, so to find the answer, divide:

$$
\begin{array}{r}
12 \\
6\overline{)72} \\
\underline{6} \\
12 \\
\underline{12} \\
0
\end{array}
$$

Each box would have 12 cameras in it.

 ## Still Struggling

When you do a word problem, always make sure that the answer you get sounds reasonable. This will help you in determining whether or not you made a mistake. For example, if you are asked to find the reduced price of an item that costs $50, and you get $75, you know this is a mistake because the reduced price of the item would be more than the original price.

Practice

Solve:

1. Find the total number of passes completed for the following players: Montana, 3,409; Lomax, 1,817; Anderson, 2,654; and White, 1,761.

2. Ieisha paid $675 for his new computer system. Included in the price was the printer at a cost of $83. How much would the system have cost without the printer?

3. Lunch for 8 people costs $112. If they decided to split the cost equally, how much would each person pay?

4. A real estate developer bought 43 acres of land at $2,750 per acre. What was the total cost of the land?

5. Mark purchased 12 roses for $36. How much was each rose for?

Answers

1. 9,641 passes completed
2. $592
3. $14
4. $118,250
5. $3

In this chapter, the basic operations of addition, subtraction, multiplication, and division of whole numbers were explained.

QUIZ

1. **Name 32,321.**
   A. three ten thousands, two thousands, three hundreds, two tens, and one ones
   B. thirty-two thousand, three hundred twenty-one
   C. thirty-two million, three hundred twenty-one
   D. thirty-two hundred, three hundred twenty-one

2. **Name 50,000,002.**
   A. fifty million two
   B. fifty billion two
   C. fifty-two million
   D. fifty-two billion

3. **Round 6,314,259 to the nearest hundred thousand.**
   A. 6,000,000
   B. 6,310,000
   C. 6,314,000
   D. 6,300,000

4. **Round 789,961 to the nearest hundred.**
   A. 790,000
   B. 789,600
   C. 800,000
   D. 789,000

5. **Round 2,867 to the nearest ten.**
   A. 2,900
   B. 2,960
   C. 3,000
   D. 2,870

6. **Add 97 + 148 + 6 + 40.**
   A. 291
   B. 281
   C. 203
   D. 213

7. **Add 17,432 + 627,381 + 62.**
   A. 645,433
   B. 645,343
   C. 644,875
   D. 644,785

8. Add 64,155 + 28,006 + 571 + 8,013.
   A. 100,745
   B. 68,345
   C. 93,713
   D. 75,640

9. Subtract 800 – 348.
   A. 548
   B. 1,148
   C. 416
   D. 452

10. Subtract 656,371 – 63,285.
    A. 593,186
    B. 602,086
    C. 593,086
    D. 593,076

11. Subtract 562,371,206 – 47,500,312.
    A. 524,871,894
    B. 514,870,894
    C. 514,871,894
    D. 524,872,894

12. Multiply 97 × 43.
    A. 4,271
    B. 4,171
    C. 1,181
    D. 1,273

13. Multiply 5,162 × 326.
    A. 1,682,812
    B. 1,692,812
    C. 1,692,912
    D. 1,682,822

14. Multiply 4,005 × 207.
    A. 800,035
    B. 830,035
    C. 829,135
    D. 829,035

15. Divide 3,796 ÷ 73.
   A. 25
   B. 61
   C. 52
   D. 35 R3

16. Divide 38,828 ÷ 571.
   A. 66
   B. 68
   C. 63
   D. 66 R24

17. Divide 12,385 ÷ 152.
   A. 81 R73
   B. 81
   C. 81 R37
   D. 82

18. In a 30-minute television program, there were 12 one-minute commercials. How many minutes of actual programming were there?
   A. 42 minutes
   B. 18 minutes
   C. 260 minutes
   D. 28 minutes

19. Tonysha bought six items costing $12, $43, $6, $18, $2, and $38. How much did she spend?
   A. $117
   B. $113
   C. $115
   D. $119

20. Pete's Pool Hall purchased 16 new billiard cues costing $11 each. What was the total cost?
   A. $166
   B. $196
   C. $176
   D. $186

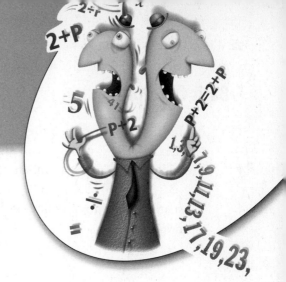

chapter **2**

Integers

This chapter explains the basic concepts of positive and negative numbers. The negative numbers are used in many areas. For example, when a person deposits $50 in a savings account, the number can be denoted as +$50. If a person withdraws $28 from the account, this number can be denoted as –$28. If the temperature is above zero, it is denoted with a positive number. If the temperature drops below zero, it is denoted by a negative number. There are many other uses of positive and negative numbers, as you will see in this chapter.

CHAPTER OBJECTIVES

In this chapter, you will learn how to

- Find the opposite and absolute value of integers
- Order integers
- Add, subtract, multiply, and divide integers
- Use exponents
- Use the order of operations

Basic Concepts

In Chapter 1, we used the set of whole numbers which consists of the numbers 0, 1, 2, 3, 4, 5, ... In algebra we extend the set of **whole numbers** by adding the negative numbers –1, –2, –3, –4, –5, ... The numbers . . . –5, –4, –3, –2, –1, 0, 1, 2, 3, 4, 5, . . . are called **integers**. These numbers can be represented on the number line, as shown in Fig. 2-1. The number zero is called the **origin**.

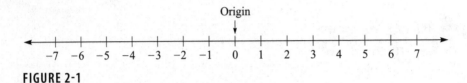

FIGURE 2-1

MATH NOTE   *Any number written without a sign (except 0) is considered to be positive. That is, 6 = +6. The number zero is neither positive nor negative.*

Each integer has an **opposite**. The opposite of a given integer is the corresponding integer which is exactly the same distance from the origin as the given integer, but in the other direction. For example, the opposite of –4 is 4. The opposite of 0 is 0.

The positive distance of any number from 0 is called the **absolute value** of the number. The symbol for absolute value is | |. Hence, |–6| = 6 and |10| = 10. In other words, the absolute value of any number except 0 is positive. The absolute value of 0 is 0. That is, |0| = 0.

Still Struggling

Do not confuse the concepts of opposite and absolute value. With the exception of zero, to find the opposite of an integer, change its sign and to find the absolute value of an integer, make it positive.

EXAMPLE

Find the opposite of 14.

SOLUTION

The opposite of 14 is –14 since we change the sign.

EXAMPLE

Find |14|.

SOLUTION

|14| = 14 since the absolute value of this number is 14.

EXAMPLE

Find the opposite of –7.

SOLUTION

The opposite of –7 is 7 since we change the sign.

EXAMPLE

Find |–7|.

SOLUTION

|–7| = 7 since the absolute value is positive.

Sometimes a negative sign is placed outside a number in parentheses. In this case, it means the opposite of the number inside the parentheses. For example, –(–6) means the opposite of –6, which is 6. Hence, –(–6) = 6. Also, –(8) means the opposite of 8, which is –8. Hence, –(8) = –8.

EXAMPLE

Find the value of –(18).

SOLUTION

The opposite of 18 is –18. Hence, –(18) = –18.

EXAMPLE

Find the value of –(–13).

SOLUTION

The opposite of –13 is 13. Hence, –(–13) = 13.

Practice

1. Find the opposite of –24.

2. Find the opposite of 11.

3. Find |–19|.

4. Find |22|.

5. Find the opposite of 0.

6. Find |0|.

7. Find the value of –(–56).

8. Find the value of –(32).

9. Find –(0).

10. Find the value of –|–7|. (Be careful.)

Answers

1. 24

2. –11

3. 19

4. 22

5. 0

6. 0

7. 56

8. –32

9. 0

10. –|–7| = –(7) = –7

Order

When comparing numbers, the symbol > means "greater than." For example, 12 > 3 is read "twelve is greater than three." The symbol < means "less than." For example, 4 < 10 is read "four is less than ten."

Still Struggling

Remember when you use the inequality signs < and >, always point the sign to the smaller number. Think of the signs as ice cream cones. The top of the cone holds more ice cream than the bottom of the cone.

Given two integers, the number further to the right on the number line is the larger number.

EXAMPLE

Compare –4 with –1.

SOLUTION

Since –1 is further to the right on the number line, it is larger than –4. See Fig. 2-2. Hence, –4 < –1 or –1 > –4.

FIGURE 2-2

EXAMPLE

Use > or < to make a true statement.

0 __ –6

SOLUTION

0 > –6, since 0 is further to the right on the number line.

Practice

Use > or < to make each a true statement.

1. 3 _____ 7
2. –9 _____ –5
3. –7 _____ 0
4. 5 _____ –6
5. –15 _____ –21

Answers

1. <
2. <
3. <
4. >
5. >

Addition of Integers

There are two basic rules for adding integers:

Rule 1: To add two integers with like signs (i.e., both integers are positive or both integers are negative), add the absolute values of the numbers and give the sum the common sign.

EXAMPLE
Add 2 + 4.

SOLUTION
Since both integers are positive, add the absolute values of each: 2 + 4 = 6; then the answer will be positive. Hence, 2 + 4 = 6.

EXAMPLE
Add (–3) + (–2).

SOLUTION
Since both integers are negative, add the absolute values: 3 + 2 = 5; then give the answer a – sign. Hence, (–3) + (–2) = –5.

The rule can be demonstrated by looking at the number lines shown in Fig. 2-3.

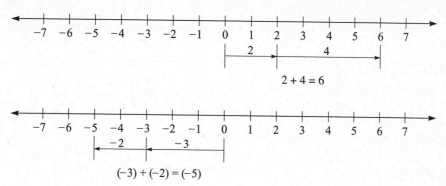

FIGURE 2-3

In the first example, you start at 0 and move 2 units to the right, ending at 2. Then from 2, move 4 units to the right, ending at 6. Therefore, 2 + 4 = 6.

In the second example, start at 0 and move 3 units to the left, ending on –3. Then from –3, move 2 units to the left, ending at –5. Therefore, (–3) + (–2) = –5.

Rule 2: To add two numbers with unlike signs (i.e., one is positive and one is negative), subtract the absolute values of the numbers and give the answer the sign of the number with the larger absolute value.

EXAMPLE
Add 5 + (–2).

SOLUTION
Since the numbers have different signs, subtract the absolute values of the numbers: 5 – 2 = 3. Then give the 3 a positive sign since 5 is larger than 3 and the sign of the 5 is positive. Therefore, 5 + (–2) = 3.

EXAMPLE
Add 3 + (–4).

SOLUTION
Since the numbers have different signs, subtract the absolute values of the numbers: 4 – 3 = 1. Then give the 1 a negative sign since 4 is larger than 3 and the sign of the 4 is negative. Therefore, 3 + (–4) = –1.

This rule can be demonstrated by looking at the number lines shown in Fig. 2-4.

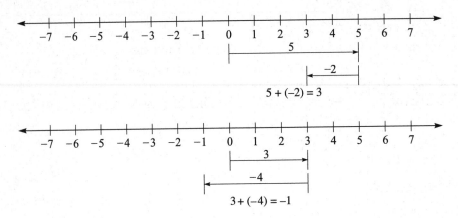

FIGURE 2-4

In the first case, start at 0 and move 5 units to the right ending on 5. From there, move 2 units to the left. You will end up at 3. Therefore, 5 + (–2) = 3.

In the second case, start at 0 and move 3 units to the right ending at 3. From there, move 4 units to the left. You will end on –1. Therefore, 3 + (–4) = –1.

To add three or more integers, you can add two at a time from left to right.

EXAMPLE

Add (–6) + (–4) + 2 + (–7) + 8.

SOLUTION

$$(-6) + (-4) + 2 + (-7) + 8 = (-10) + 2 + (-7) + 8$$
$$= (-8) + (-7) + 8$$
$$= (-15) + 8$$
$$= (-7)$$

Another way to add integers is to add the positive numbers, then add the negative numbers, and then add the answers algebraically. Remember that when we add positive and negative numbers, there are **two** rules. When the signs are alike, we add the numbers, and when the signs are unlike, we subtract the numbers.

■ **EXAMPLE**

Add $(-3) + (-4) + 2 + 8 + (-5)$.

✔ **SOLUTION**

Add $2 + 8 = 10$ and $(-3) + (-4) + (-5) = -12$. Then add $10 + (-12) = -2$.

MATH NOTE *When zero is added to any number, the answer is the number. For example, $0 + 6 = 6$, $(-3) + 0 = -3$.*

Practice
Add:

  1. $5 + 13$

  2. $(-8) + (-3)$

  3. $4 + (-7)$

  4. $9 + (-5)$

  5. $(-18) + 0$

  6. $5 + (-12) + 1$

  7. $18 + (-12) + (-6)$

  8. $9 + (-13) + 3$

  9. $(-14) + (-6) + (-2) + 9$

  10. $(-3) + 14 + 3 + (-6) + (-2)$

Answers

  1. 18

  2. –11

  3. –3

  4. 4

  5. –18

  6. –6

  7. 0

  8. –1

  9. –13

  10. 6

Subtraction of Integers

In arithmetic, subtraction is usually thought of as "taking away." For example, if you have six books on your desk at home and you take four to class, you have two books left on your desk. The "taking away" concepts work well in arithmetic, but with algebra, a new way of thinking about subtraction is necessary.

In algebra, we think of subtraction as adding the opposite. For example, in arithmetic, $8 - 6 = 2$. In algebra, $8 + (-6) = 2$. Notice that in arithmetic, we subtract 6 from 8. In algebra, we add the opposite of 6, which is -6, to 8. In both cases, we get the same answer.

To subtract one number from another, add the opposite of the number that is being subtracted.

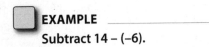 **EXAMPLE**

Subtract $14 - (-6)$.

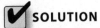 **SOLUTION**

Add 6 (the opposite of -6) to 14 to get 20 as shown.

$14 - (-6) = 14 + 6 \leftarrow$ (opposite of -6)

$\qquad = 20$

Hence, $14 - (-6) = 20$.

EXAMPLE

Subtract $(-5) - 2$.

SOLUTION

$(-5) - 2 = (-5) + (-2) \leftarrow$ (opposite of 2)

$\qquad = -7$

Hence, $(-5) - 2 = -7$.

Still Struggling

Sometimes the answers in subtraction do not look correct, but once you get the opposite, you follow the rules of addition. If the two numbers have like signs, use Rule 1 for addition. If the two numbers have unlike signs, use Rule 2 for addition.

 EXAMPLE

Subtract (−11) − (−9).

SOLUTION

(−11) − (−9) = (−11) + 9 ← (opposite of −9)

= −2

Hence, (−11) − (−9) = −2.

 EXAMPLE

Subtract (−7) − (−15).

SOLUTION

(−7) − (−15) = (−7) + 15 ← (opposite of −15)

= 8

Hence, (−7) − (−15) = 8.

Practice

Subtract:

1. 12 − 9

2. (−14) − (−13)

3. (−6) − 19

4. 17 − (−21)

5. $(-4) - (-15)$
6. $(-11) - (-3)$
7. $(-17) - (-17)$
8. $16 - (-9)$
9. $(-8) - 8$
10. $0 - (-5)$

Answers

1. 3
2. –1
3. –25
4. 38
5. 11
6. –8
7. 0
8. 25
9. –16
10. 5

Addition and Subtraction

When performing the operations of addition and subtraction in the same problem, follow these steps:

- Step 1: Change all the subtractions to addition. Remember to add the opposite.
- Step 2: Add left to right.

EXAMPLE

Perform the indicated operations:

$5 + (-8) - (-3) + 6 - 11 + 7 - (-4)$

✔ **SOLUTION**

Step 1: $5 + (-8) + 3 + 6 + (-11) + 7 + 4$

Step 2: $-3 + 3 + 6 + (-11) + 7 + 4$

$= 0 + 6 + (-11) + 7 + 4$

$= 6 + (-11) + 7 + 4$

$= -5 + 7 + 4$

$= 2 + 4$

$= 6$

Hence, the answer is 6.

Practice

For each, perform the indicated operations.

1. $-8 + 2 - 9$
2. $16 - (-4) - 7$
3. $-15 + 6 - (-2) - (-9)$
4. $5 + (-10) - (-7)$
5. $-3 + 7 + 4 - 9 - 2 - 5$

Answers

1. -15
2. 13
3. 2
4. 2
5. -8

Multiplication of Integers

For multiplication of integers, there are two basic rules:

Rule 1: To multiply two integers with the same signs (i.e., both are positive or both are negative), multiply the absolute values of the numbers and give the answer a + (positive) sign.

EXAMPLE

Multiply 6×3.

SOLUTION

Multiply $6 \times 3 = 18$. Since both integers are positive, give the answer a + (positive) sign. Hence, $6 \times 3 = 18$.

EXAMPLE

Multiply $(-7) \times (-4)$.

SOLUTION

Multiply $7 \times 4 = 28$. Since both integers are negative, the answer is positive. Hence, $(-7) \times (-4) = 28$.

Rule 2: To multiply two integers with unlike signs (i.e., one integer is positive and one integer is negative), multiply the absolute values of the numbers and give the answer a – (negative) sign.

EXAMPLE

Multiply $(-9) \times 4$.

SOLUTION

Multiply $9 \times 4 = 36$, and give the answer a – (negative) sign. Hence, $(-9) \times 4 = -36$.

EXAMPLE

Multiply $8 \times (-7)$

SOLUTION

Multiply $8 \times 7 = 56$ and give the answer a – (negative) sign. Hence, $8 \times (-7) = -56$.

MATH NOTE *Multiplication can be shown without a times sign. For example, $(-3)(-5)$ means $(-3) \times (-5)$. Also, a dot can be used to represent multiplication. For example, $5 \cdot 3 \cdot 2$ means $5 \times 3 \times 2$.*

To multiply three or more nonzero integers, multiply the absolute values and count the number of negative numbers. If there is an odd number of negative numbers, give the answer a – (negative) sign. If there is an even number of negative numbers, give the answer a + (positive) sign.

EXAMPLE

Multiply (–7)(–3)(–4).

SOLUTION

Multiply 7 × 3 × 4 = 84. Since there are 3 negative numbers, the answer is negative. Hence, (–7)(–3)(–4) = –84.

EXAMPLE

Multiply (–3)(–6)(2)(8).

SOLUTION

Multiply 3 × 6 × 2 × 8 = 288. Since there are two negative numbers, the answer is positive. Hence, (–3)(–6)(2)(8) = +288.

Practice
Multiply:

1. (–9)(–6)
2. (7)(–3)
3. (6)(5)
4. (–4)(7)
5. (8)(–5)
6. (4)(–3)(–6)
7. (7)(–2)(9)(–4)
8. (15)(–9)(6)(2)
9. (–3)(–12)(4)(–6)(–4)
10. (–6)(4)(–2)(–3)(–8)(6)(–1)

Answers

1. 54
2. –21
3. 30
4. –28
5. –40
6. 72
7. 504
8. –1620
9. 3456
10. –6912

Division of Integers

Division can be represented in three ways:

1. The division box

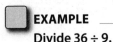

2. The division sign $16 \div 8 = 2$

3. Fraction notation sign $\dfrac{16}{8} = 2$

The rules for division of integers are the same as the rules for multiplication of integers.

Rule 1: To divide two integers with like signs, divide the absolute values of the numbers and give the answer a + sign.

EXAMPLE

Divide 36 ÷ 9.

SOLUTION

Divide 36 ÷ 9 = 4. Since both integers are positive, give the answer a + sign.
Hence, 36 ÷ 9 = 4.

EXAMPLE

Divide (–24) ÷ (–3).

 SOLUTION

Divide 24 ÷ 3 = 8. Since both integers are negative, the answer is positive.
Hence, (–24) ÷ (–3) = 8.

*Rule 2: To divide two integers with unlike signs, divide the absolute values of the
integers and give the answer a – sign.*

 EXAMPLE

Divide (–45) ÷ 5.

 SOLUTION

Divide 45 ÷ 5 = 9. Since the numbers have unlike signs, give the answer
a – sign. Hence, (–45) ÷ 5 = –9.

EXAMPLE

Divide 21 ÷ (–7).

SOLUTION

Divide 21 ÷ 7 = 3. Since the numbers have unlike signs, give the answer
a – sign. Hence, 21 ÷ (–7) = –3.

MATH NOTE *When division is written as a fraction and one of the numbers is
negative, the negative sign can be written with the top number or the bottom
number or in front of the fraction. For example,* $\dfrac{-6}{3} = \dfrac{6}{-3} = -\dfrac{6}{3} = -2.$

Practice
Divide:

1. (–72) ÷ (–9)
2. 48 ÷ (–6)
3. 21 ÷ 7

4. $(-60) \div 5$

5. $(-74) \div (-2)$

Answers

1. 8

2. –8

3. 3

4. –12

5. 37

Exponents

When the same number is multiplied by itself, the indicated product can be written in **exponential notation**. For example, 3×3 can be written as 3^2, where the 3 is called the **base** and the 2 is called the **exponent**. Also,

$3 \times 3 \times 3 = 3^3$
$3 \times 3 \times 3 \times 3 = 3^4$
$3 \times 3 \times 3 \times 3 \times 3 = 3^5$, etc.

3^2 is read as "three squared" or "three to the second power." 3^3 is read as "three cubed" or "three to the third power." 3^4 is read as "three to the fourth power," etc.

MATH NOTE *When no exponent is written with a number, it is assumed to be one. For example, $3 = 3^1$.*

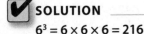 **EXAMPLE**
Find 6^3.

 SOLUTION
$6^3 = 6 \times 6 \times 6 = 216$

EXAMPLE
Find 2^7.

SOLUTION
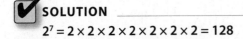 $2^7 = 2 \times 2 \times 2 \times 2 \times 2 \times 2 \times 2 = 128$

Exponents can be used with negative numbers as well. For example, $(-8)^3$ means $(-8) \times (-8) \times (-8)$. Notice that in the case of negative numbers, the integer must be enclosed in parentheses. When the $-$ sign is **not** enclosed in parentheses, it is **not** raised to the power. For example, -8^3 means $-(8 \cdot 8 \cdot 8)$.

EXAMPLE

Find $(-4)^6$.

SOLUTION

$(-4)^6 = (-4)(-4)(-4)(-4)(-4)(-4) = 4{,}096$

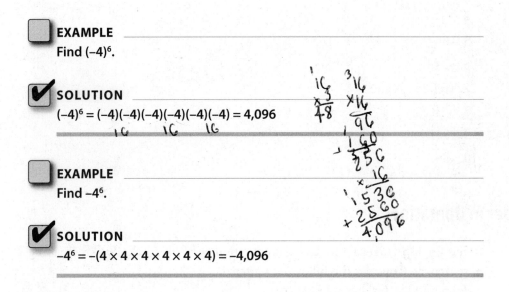

EXAMPLE

Find -4^6.

SOLUTION

$-4^6 = -(4 \times 4 \times 4 \times 4 \times 4 \times 4) = -4{,}096$

MATH NOTE   *The number one raised to any power will always be one. For example, $1^4 = 1 \times 1 \times 1 \times 1 = 1$ and $1^{32} = 1$.*

Practice
Find each of the following:

1. 5^4

2. 2^9

3. 6^2

4. 3^1

5. $(-4)^3$

6. $(-3)^5$

7. $(-5)^6$

8. $(-8)^2$

9. (-6^3)

10. (-7^3)

Answers

1. 625
2. 512
3. 36
4. 3
5. −64
6. −243
7. 15,625
8. 64
9. −216
10. −343

Order of Operations

In the English language, we have punctuation symbols to clarify the meaning of sentences. Consider the following sentence:

John said the teacher is tall.

This sentence could have two different meanings depending on how it is punctuated:

John said, "The teacher is tall."

or

"John," said the teacher, "is tall."

In mathematics, we have what is called an **order of operations** to clarify the meaning when there are operations and grouping symbols (parentheses) in the same problem.

The order of operations is

1. Parentheses

2. Exponents

3. Multiplication or Division, left to right

4. Addition or Subtraction, left to right

Multiplication and division are equal in order and should be performed from left to right. Addition and subtraction are equal in order and should be performed from left to right.

MATH NOTE   *The word simplify means to perform the operations following the order of operations.*

EXAMPLE

Simplify $24 - 6 \times 2 - 10 \div 2$.

SOLUTION

$$24 - 6 \times 2 - 10 \div 2 = 24 - 12 - 5 \quad \text{multiplication and division, left to right}$$
$$= 7 \quad \text{subtraction left to right}$$

EXAMPLE

Simplify $6 + 3^2 - 2 \times 8$.

SOLUTION

$$6 + 3^2 - 2 \times 8 = 6 + 9 - 2 \times 8 \quad \text{exponent}$$
$$= 6 + 9 - 16 \quad \text{multiplication}$$
$$= 15 - 16 \quad \text{addition}$$
$$= -1 \quad \text{subtraction}$$

EXAMPLE

Simplify $59 - (10 - 4) \times 2^3$.

SOLUTION

$$59 - (10 - 4) \times 2^3 = 59 - 6 \times 2^3 \quad \text{parentheses}$$
$$= 59 - 6 \times 8 \quad \text{exponent}$$
$$= 59 - 48 \quad \text{multiplication}$$
$$= 11 \quad \text{subtraction}$$

EXAMPLE

Simplify $23 + (8 - 4)^2 - 18 \div 2$.

SOLUTION

$$23 + (8 - 4)^2 - 18 \div 2 = 23 + (4)^2 - 18 \div 2 \quad \text{parentheses}$$
$$= 23 + 16 - 18 \div 2 \quad \text{exponent}$$
$$= 23 + 16 - 9 \quad \text{division}$$
$$= 39 - 9 \quad \text{addition}$$
$$= 30 \quad \text{subtraction}$$

Still Struggling

The order of operations, Parentheses, Exponents, Multiplication, Division, Addition, Subtraction, can be remembered by making up a sentence using the first letter of each word (PEMDAS)—for example, Please Excuse My Dear Aunt Sally. Remember multiplication and division are equal in order and are performed from left to right. Also, addition and subtraction are equal in order and are performed left to right.

Practice

Simplify:

1. $6 - 3^2 + 5$
2. $24 \div 8 \times 3$
3. $45 - (7 \times 2)^2 + 18 \div 6$
4. $9^2 - 4 \times 15 \div 3$
5. $6^3 + (2 \times 5^2 - 10) \div 4$

Answers

1. 2
2. 9
3. −148
4. 61
5. 226

This chapter explained the basic operations using the integers. Also, exponents and the order of operations were explained.

QUIZ

1. Find |–14|.
   A. 14
   B. –14
   C. 0
   D. |4|

   (A)

2. Find the opposite of 27.
   A. |27|
   B. –27
   C. 0
   D. 27

   (B)

3. Find the value of –(–13).
   A. –13
   B. 0
   C. –(+13)
   D. 13

   (A)

4. Which number is the largest: –5, 0, 2, –3?
   A. 0
   B. –5
   C. 2
   D. –3

   (C)

5. Add –14 + 8.
   A. –22
   B. 6
   C. –6
   D. 22

   (C) $\begin{array}{r} 14 \\ -8 \\ \hline 6 \end{array}$

6. Add –7 + (–12).
   A. –19
   B. 5
   C. –5
   D. 19

   (A)

7. Add 12 + (–5) + (–8) + (–6).
   A. 31
   B. –2
   C. 21
   D. –7

   (D) $7 + -14$

   -7

8. Subtract −15 − 7.
   A. −8   KCC
   B. −22   −15 + 7
   C. 8   −8   Ⓐ or Ⓓ
   D. 22

9. Subtract −6 − (−10).
   A. 4   KCC
   B. −16   −6 + 10
   C. 16   4   Ⓐ
   D. −4

10. Subtract 17 − (−9).
    A. 8   KCC
    B. −26   17 + 9
    C. −8   26   Ⓓ
    D. 26

11. Simplify 15 − 7 + 5 + 8 − 6.
    A. 5   12 + 8   5 + 8 −
    B. 25   15−7   20 −6   15 − 7 + 13 − 6
    Ⓒ 15   8 +   Ⓒ   8 + 13 − 6
    D. −5   21 − 6 = 15

12. Multiply (−7)(8).
    A. 56   56
    B. −15   Ⓐ
    C. −56
    D. 15   Ⓒ

13. Multiply (−16)(−4).
    A. −20   ¹6
    B. 64   × 4
    C. 4   Ⓑ 64
    D. −64

14. Multiply (−12)(−2)(6)(3)(−5).
    A. −2,160   24 • 18 • −5   ⁴18   ³24
    B. 216   × 5   × 90
    C. −216   24 • 90   90   00
    D. 2,160   + 2160
                        Ⓓ   2160

15. Divide (−64) ÷ 4.

A. 16

B. −16

C. 60

D. −216

16. Divide (−33) ÷ (−11).

A. −3

B. 11

C. 3

D. −11

17. Find (−8)³.

A. −512

B. −24

C. 24

D. 512

18. Simplify 12 − 8 × 2 + 4.

A. 12

B. 10

C. −36

D. 0

19. Simplify 7 × 2³ + 6 − 1.

A. 47

B. 61

C. 91

D. 96

20. Simplify 4 × (8 − 3 × 2) ÷ 2.

A. 20

B. 5

C. 4

D. 16

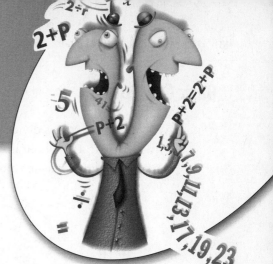

chapter 3

Fractions: Part 1

This chapter explains some basic concepts of fractions. In it you will learn how to reduce fractions, change fractions to higher terms, convert improper fractions to mixed numbers, and mixed numbers to improper fractions. Fractions have many uses in our daily life. For example, food preparations use fractions, such as $\frac{1}{4}$ cup of flour. Patterns for making clothes use fractions, such as $3\frac{1}{2}$ yards of material. Budgets use fractions, medical prescriptions use fractions, and so on.

CHAPTER OBJECTIVES

In this chapter, you will learn how to

- Reduce a fraction
- Change a fraction to higher terms
- Change an improper fraction to a mixed number
- Change a mixed number to an improper fraction

Basic Concepts

When a whole item is divided into equal parts, the relationship of one or more of the parts to the whole is called a **fraction**. When a pie is cut into six equal parts, each part is one-sixth of the whole pie, as shown in Fig. 3-1.

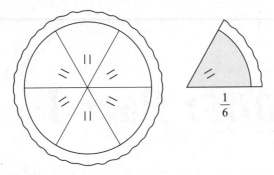

FIGURE 3-1

The symbol for a fraction is two numbers separated by a bar. The fraction five-eighths is written as $\frac{5}{8}$. The top number of the fraction is called the **numerator** and the bottom number is called the **denominator**.

$$\frac{5}{8} \quad \begin{matrix} \leftarrow \text{ numerator} \\ \leftarrow \text{ denominator} \end{matrix}$$

The denominator tells how many parts the whole is being divided into, and the numerator tells how many parts are being used. The fraction $\frac{5}{8}$ means five equal parts of the whole that has been divided into eight equal parts.

MATH NOTE   *The denominator of a fraction cannot be zero.*

Other fractional parts are shown in Fig. 3-2.

Still Struggling

Fractions have two meanings. For example, $\frac{3}{4}$ means 3 parts out of a total of 4 parts. It could also mean $3 \div 4$ or 0.75.

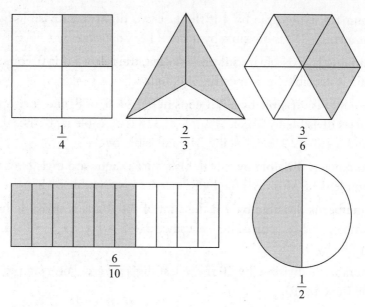

FIGURE 3-2

A fraction whose numerator is less than its denominator is called a **proper** fraction. For example, $\frac{5}{8}, \frac{2}{3}$, and $\frac{1}{6}$ are proper fractions. A fraction whose numerator is greater than or equal to its denominator is called an **improper** fraction. For example, $\frac{5}{3}, \frac{6}{6}$, and $\frac{10}{4}$ are improper fractions. A number that consists of a whole number and a fraction is called a **mixed** number. For example, $6\frac{5}{9}, 1\frac{2}{3}$, and $3\frac{1}{8}$ are mixed numbers.

A mixed number, such as $2\frac{3}{4}$, means a whole number, 2, plus a fraction, $\frac{3}{4}$. For example, $2\frac{3}{4} = 2 + \frac{3}{4}$.

Divisibility Rules

When reducing fractions, it is necessary to find a number which divides into the numerator and denominator of the fractions. Sometimes this may be difficult, so you can use what are called the divisibility rules.

- **A number is divisible by 2 if the last digit is 0, 2, 4, 6, or 8.** For example, 326 is divisible by 2 since the last digit is 6 and 6 is divisible by 2.

- **A number is divisible by 3 if the sum of the digits is divisible by 3.** For example, 537 is divisible by 3 since 5 + 3 + 7 = 15 and 15 is divisible by 3.

- A number is divisible by 4 if the last two digits are divisible by 4. For example, 13,812 is divisible by 4 since 12 is divisible by 4.
- A number is divisible by 5 if the last digit is either a 5 or a 0. For example, 8,365 is divisible by 5 since the last digit is 5.
- A number is divisible by 6 if it ends in 0, 2, 4, 6, or 8, and the sum of the digits is divisible by 3. For example, 8,328 is divisible by 6 since it ends in 8, and 8 + 3 + 2 + 8 = 21, and 21 is divisible by 3.
- A number is divisible by 8 if the last three digits are divisible by 8. For example, 13,824 is divisible by 8 since 824 is divisible by 8.
- A number is divisible by 9 if the sum of the digits is divisible by 9. For example, 35,118 is divisible by 9 since 3 + 5 + 1 + 1 + 8 = 18, and 18 is divisible by 9.
- A number is divisible by 10 if the last digit is zero. For example, 16,240 is divisible by 10.

These rules should help you when you have to reduce a fraction.

Reducing Fractions

A fraction is said to be in **lowest terms** if both the numerator and denominator cannot be divided evenly by any number except one.

To reduce a fraction to lowest terms, divide the numerator and the denominator by the largest number that divides evenly into both.

EXAMPLE

Reduce $\dfrac{21}{28}$ to lowest terms.

SOLUTION

Divide both numerator and denominator by 7, as shown:

$$\frac{21}{28} = \frac{21 \div 7}{28 \div 7} = \frac{3}{4}$$

EXAMPLE

Reduce $\dfrac{10}{25}$ to lowest terms.

SOLUTION

Divide the numerator and denominator by 5, as shown:

$$\frac{10}{25} = \frac{10 \div 5}{25 \div 5} = \frac{2}{5}$$

If the largest number that divides evenly into both numerator and denominator is not obvious, divide the numerator and denominator by any number (except one) that divides into each evenly, then repeat the process until the fraction is in lowest terms.

EXAMPLE

Reduce $\dfrac{84}{147}$ to lowest terms.

SOLUTION

First divide by 3:

$$\frac{84}{147} = \frac{84 \div 3}{147 \div 3} = \frac{28}{49}$$

Next divide by 7:

$$\frac{28}{49} = \frac{28 \div 7}{49 \div 7} = \frac{4}{7}$$

MATH NOTE   *When the numerator of a fraction is zero, the value of the fraction is zero. For example, $\dfrac{0}{6} = 0$.*

Practice

Reduce each fraction to lowest terms.

1. $\dfrac{6}{15}$

2. $\dfrac{5}{20}$

3. $\dfrac{20}{36}$

4. $\dfrac{9}{21}$

5. $\dfrac{150}{400}$

6. $\dfrac{48}{64}$

7. $\dfrac{35}{60}$

8. $\dfrac{150}{216}$

9. $\dfrac{68}{119}$

10. $\dfrac{6,000}{20,000}$

Answers

1. $\dfrac{2}{5}$

2. $\dfrac{1}{4}$

3. $\dfrac{5}{9}$

4. $\dfrac{3}{7}$

5. $\dfrac{3}{8}$

6. $\dfrac{3}{4}$

7. $\dfrac{7}{12}$

8. $\dfrac{25}{36}$

9. $\dfrac{4}{7}$

10. $\dfrac{3}{10}$

Changing Fractions to Higher Terms

The opposite of reducing fractions is changing fractions to higher terms.

To change a fraction to an equivalent fraction with a larger denominator, divide the larger denominator by the smaller denominator and then multiply the numerator of the fraction with the smaller denominator by the number obtained to get the numerator of the fraction with the larger denominator.

 EXAMPLE

Change $\dfrac{2}{5}$ into an equivalent fraction with a denominator of 15.

 SOLUTION

Divide 15 by 5 to get 3, then multiply 3 by 2 to get 6.

Hence, $\dfrac{2}{5} = \dfrac{2 \times 3}{5 \times 3} = \dfrac{6}{15}$

 EXAMPLE

Change $\dfrac{7}{12}$ to an equivalent fraction with a denominator of 48.

 SOLUTION

$$\frac{7}{12} = \frac{7 \times 4}{12 \times 4} = \frac{28}{48}$$

Practice

1. Change $\dfrac{1}{11}$ to 44ths.

2. Change $\dfrac{5}{9}$ to 54ths.

3. Change $\dfrac{7}{16}$ to 80ths.

4. Change $\dfrac{31}{36}$ to 252nds.

5. Change $\dfrac{7}{10}$ to 80ths.

Answers

1. $\dfrac{4}{44}$

2. $\dfrac{30}{54}$

3. $\dfrac{35}{80}$

4. $\dfrac{217}{252}$

5. $\dfrac{56}{80}$

Changing Improper Fractions to Mixed Numbers

An improper fraction can be changed into an equivalent mixed number. For example, $\dfrac{15}{4}$ is the same as $3\dfrac{3}{4}$.

To change an improper fraction to an equivalent mixed number, first divide the numerator by the denominator. Write the answer as a whole number and the remainder as a fraction with the divisor as the denominator and the remainder as the numerator. It may be necessary to reduce the fraction.

 EXAMPLE

Change $\dfrac{18}{7}$ to a mixed number.

 SOLUTION

Divide 18 by 7, as shown:

$$\begin{array}{r} 2 \\ 7\overline{)18} \\ -14 \\ \hline 4 \end{array}$$

Write the answer as $2\dfrac{4}{7}$.

 EXAMPLE

Change $\dfrac{28}{16}$ to a mixed number.

SOLUTION

Divide 28 by 16, as shown:

$$16\overline{)28}$$

with quotient 1, subtracting 16 leaving 12.

$$
\begin{array}{r}
1 \\
16\overline{)28} \\
\underline{16} \\
12
\end{array}
$$

Write the answer as $1\frac{12}{16}$ and reduce $\frac{12}{16}$ to $\frac{3}{4}$. Hence, the answer is $1\frac{3}{4}$.

MATH NOTE *Any fraction where the numerator is equal to the denominator is always equal to one. For example, $\frac{7}{7} = 1$, $\frac{12}{12} = 1$, etc.*

Improper fractions are sometimes equal to whole numbers. This happens when the remainder is zero.

EXAMPLE

Change $\frac{25}{5}$ to a mixed or whole number.

 SOLUTION

Divide 25 by 5, as shown:

$$
\begin{array}{r}
5 \\
5\overline{)25} \\
\underline{25} \\
0
\end{array}
$$

Write the answer as 5.

Practice

Change each to a mixed number in lowest terms.

1. $\dfrac{11}{6}$

2. $\dfrac{7}{2}$

3. $\dfrac{23}{8}$

4. $\dfrac{36}{24}$

5. $\dfrac{50}{12}$

Answers

1. $1\dfrac{5}{6}$

2. $3\dfrac{1}{2}$

3. $2\dfrac{7}{8}$

4. $1\dfrac{1}{2}$

5. $4\dfrac{1}{6}$

Changing Mixed Numbers to Improper Fractions

A mixed number can be changed to an equivalent improper fraction. For example, $6\dfrac{3}{8}$ is equal to $\dfrac{51}{8}$.

To change a mixed number to an improper fraction, multiply the whole number by the denominator of the fraction and add the numerator of the fraction to the product. Use this number for the numerator and use the same denominator as the denominator of the improper fraction.

 EXAMPLE

Change $8\dfrac{3}{4}$ to an improper fraction.

 SOLUTION

Multiply 8 by 4, then add 3 to get 35, which is the numerator of the improper fraction. Use 4 as the denominator of the improper fraction.

$$8\dfrac{3}{4} = \dfrac{4 \times 8 + 3}{4} = \dfrac{35}{4}$$

 EXAMPLE

Change $6\dfrac{3}{5}$ to an improper fraction.

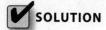 SOLUTION

$$6\frac{3}{5} = \frac{5 \times 6 + 3}{5} = \frac{33}{5}$$

MATH NOTE *Any whole number can be written as a fraction by placing the whole number over one. For example, $7 = \frac{7}{1}$, $13 = \frac{13}{1}$, etc. Likewise, any fraction with a denominator of one can be written as a whole number $\frac{6}{1} = 6$, $\frac{10}{1} = 10$, etc.*

Practice

Change each to an improper fraction.

1. $8\frac{9}{11}$

2. $7\frac{2}{3}$

3. $6\frac{5}{8}$

4. $3\frac{9}{10}$

5. $12\frac{2}{3}$

Answers

1. $\frac{97}{11}$

2. $\frac{23}{3}$

3. $\frac{53}{8}$

4. $\frac{39}{10}$

5. $\frac{38}{3}$

This chapter explained the basic concepts of fractions and mixed numbers. Fractions can be reduced or changed to higher terms. Improper fractions can be changed to mixed numbers and mixed numbers can be changed to improper fractions.

QUIZ

1. What fraction is shown by the shaded portion of Fig. 3-3?

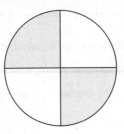

FIGURE 3-3

A. $\dfrac{1}{2}$

B. $\dfrac{1}{3}$ $\dfrac{2}{4} \div \dfrac{2}{2} = \dfrac{1}{2}$

C. $\dfrac{1}{4}$ Ⓐ

D. $\dfrac{3}{4}$

2. What fraction is shown by the shaded portion of Fig. 3-4?

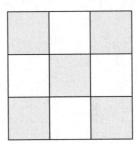

FIGURE 3-4

A. $\dfrac{4}{5}$ $\dfrac{5}{9}$

B. $\dfrac{5}{9}$ Ⓑ

C. $\dfrac{5}{4}$

D. $\dfrac{4}{9}$

3. Which of the following is a proper fraction?

 A. $\dfrac{8}{1}$

 B. $\dfrac{9}{7}$

 C. $\dfrac{2}{3}$ ⓒ

 D. $\dfrac{6}{6}$

4. Which of the following is an improper fraction?

 A. $\dfrac{11}{8}$

 B. $\dfrac{3}{4}$ Ⓐ

 C. $\dfrac{1}{5}$

 D. $\dfrac{9}{10}$

5. Which of the following is a mixed number?

 A. $\dfrac{3}{5}$

 B. $\dfrac{8}{5}$

 C. $\dfrac{6}{2}$ Ⓓ

 D. $2\dfrac{3}{8}$

6. Reduce $\dfrac{54}{72}$ to lowest terms.

 A. $\dfrac{4}{3}$

 B. $\dfrac{3}{4}$

 C. $\dfrac{5}{6}$ Ⓑ

 D. $\dfrac{5}{8}$

$$\frac{54 \div 2}{72 \div 2} = \frac{27}{36}$$

$$\frac{27 \div 3}{36 \div 3} = \frac{9 \div 3}{12 \div 3} = \frac{3}{4}$$

7. Reduce $\dfrac{15}{45}$ to lowest terms.

   A. $\dfrac{3}{1}$

   B. $\dfrac{3}{5}$

   C. $\dfrac{5}{3}$

   D. $\dfrac{1}{3}$

   $$\frac{15 \div 5}{45 \div 5} = \frac{3}{9} \div \frac{\div 3}{\div 3} = \frac{1}{3}$$

   (D)

8. Reduce $\dfrac{135}{144}$ to lowest terms.

   A. $\dfrac{15}{16}$

   B. $\dfrac{27}{38}$

   C. $\dfrac{30}{32}$

   D. $\dfrac{7}{12}$

   $$\frac{135 \div 3}{144 \div 3} = \frac{45 \div 3}{48 \div 3} = \frac{15}{16}$$

   (A)

9. Change $\dfrac{7}{8}$ to an equivalent fraction in higher terms.

   A. $\dfrac{12}{16}$

   B. $\dfrac{54}{64}$

   C. $\dfrac{21}{24}$

   D. $\dfrac{29}{32}$

   $$\frac{7 \times 2}{8 \times 2} = \frac{14}{16}$$

   $$\frac{7 \times 3}{8 \times 3} = \frac{21}{24}$$

   (C)

10. Change $\dfrac{7}{9}$ to an equivalent fraction in higher terms.

   A. $\dfrac{26}{36}$

   B. $\dfrac{15}{18}$

   C. $\dfrac{42}{54}$

   D. $\dfrac{64}{81}$

   (C)

   $$\frac{7 \times 2}{9 \times 2} = \frac{14}{18}$$

   $$\frac{7 \times 3}{9 \times 3} = \frac{21}{27}$$

   $$\frac{7 \times 6}{9 \times 6} = \frac{42}{54}$$

   $$\frac{7 \times 4}{9 \times 4} = \frac{28}{36}$$

   $$\frac{7 \times 5}{9 \times 5} = \frac{35}{45}$$

11. Change $\frac{1}{8}$ to an equivalent fraction in higher terms.

    A. $\frac{12}{32}$

    B. $\frac{8}{24}$ (D)

    C. $\frac{5}{32}$

    D. $\frac{3}{24}$

$$\frac{1\times2}{8\times2}=\frac{2}{16} \qquad \frac{1\times3}{8\times3}=\frac{3}{24} \qquad \frac{1\times4}{8\times4}=\frac{4}{32}$$

$$\frac{1\times7}{8\times7}=\frac{7}{56} \qquad \frac{1\times5}{8\times5}=\frac{5}{40} \qquad \frac{1\times6}{8\times6}=\frac{6}{48}$$

12. Change $\frac{11}{4}$ to a mixed number.

    A. $2\frac{3}{4}$

    B. $3\frac{1}{4}$

    C. $2\frac{1}{4}$

    D. $5\frac{1}{2}$ (A)

$$\frac{11}{4}$$

$$\begin{array}{r} 2.75 \\ 4\overline{)11.00} \\ -8 \\ \hline 30 \\ -28 \\ \hline 20 \end{array}$$

$$2\frac{11}{4}$$

$$\begin{array}{r} 3\frac{}{} \\ 2\overline{)18} \\ -6 \\ \hline 13 \end{array}$$

13. Change $\frac{24}{5}$ to a mixed number.

    A. $5\frac{1}{5}$

    B. $4\frac{4}{5}$

    C. $3\frac{2}{5}$

    D. $1\frac{1}{8}$ (B)

$$\frac{24}{5} \quad 4\frac{4}{5}$$

14. Change $6\frac{3}{8}$ to an improper fraction.

    A. $\frac{17}{8}$

    B. $\frac{23}{8}$

    C. $\frac{16}{8}$

    D. $\frac{51}{8}$ (D)

$$6\begin{smallmatrix}+3\\ \times8\end{smallmatrix} \qquad 48+3=\frac{51}{8}$$

15. Change $6\dfrac{3}{5}$ to an improper fraction.

A. $\dfrac{33}{5}$

B. $\dfrac{14}{5}$

C. $\dfrac{23}{3}$

D. $\dfrac{5}{14}$

$6\dfrac{+3}{\times 5}$ $30+3=\dfrac{33}{5}$

(A)

16. Change $9\dfrac{5}{12}$ to an improper fraction.

A. $\dfrac{57}{9}$

B. $\dfrac{113}{12}$

C. $\dfrac{69}{5}$

D. $\dfrac{45}{5}$

$9\dfrac{+5}{\times 12}$ $108+5=\dfrac{113}{12}$

$\begin{array}{r}12 \\ \times 9 \\ \hline 108\end{array}$

$\begin{array}{r}108 \\ +\ 5 \\ \hline 113\end{array}$ (B)

17. Write 14 as a fraction.

A. $\dfrac{14}{1}$

B. $\dfrac{1}{14}$

C. $\dfrac{14}{14}$

D. $\dfrac{28}{14}$

(A)

18. Write $\dfrac{7}{7}$ as a whole number.

A. 49

B. 14

C. 0

D. 1

(D)

19. What number is $\dfrac{0}{3}$ equal to?

A. 0

B. 3

C. 30

D. Cannot have 0 in the numerator of a fraction

(D) or (B)

20. **Which number cannot be used as the denominator of a fraction?**
    A. 1
    B. 0
    C. 6
    D. 100

chapter 4

Fractions: Part 2

This chapter explains the basic operations of addition, subtraction, multiplication, and division using fractions and mixed numbers. In addition, operations with positive and negative fractions are explained.

CHAPTER OBJECTIVES

In this chapter, you will learn how to

- Find a common denominator of fractions
- Add, subtract, multiply, and divide fractions
- Add subtract, multiply, and divide mixed numbers
- Solve word problems using fractions
- Compare fractions
- Perform operations with positive and negative fractions

Finding Common Denominators

In order to add or subtract two or more fractions, they must have the same denominator. This denominator is called a **common denominator**. For any two or more fractions, there are many common denominators; however, in mathematics, we usually use what is called the **lowest (or least) common denominator**, abbreviated as **LCD**.

Suppose you wanted to add $\frac{1}{2}$ and $\frac{2}{5}$. Since halves and fifths are different sizes, they cannot be added directly. It is necessary to convert each to equivalent fractions with the same denominator. This can be accomplished by changing each to tenths. Ten is the lowest common denominator of $\frac{1}{2}$ and $\frac{2}{5}$. Since $\frac{1}{2} = \frac{5}{10}$ and $\frac{2}{5} = \frac{4}{10}$, the two fractions can now be added as $\frac{5}{10} + \frac{4}{10} = \frac{9}{10}$.

There are several methods of finding the lowest common denominator. The easiest method is to simply look at the numbers in the denominator of the fractions and "see" what is the smallest number that all the denominator numbers divide into evenly. For example, 2 and 5 both divide into 10 evenly. However, this only works when the denominators are small numbers.

Another method is to list the multiples of the numbers in the denominators and eventually you will find a common multiple which can be used as a common denominator.

A **multiple** of a given number is the product of the given number and any other whole number. Multiples of a given number are obtained by multiplying the given number by 0, 1, 2, 3, 4, 5, etc. For example, the multiples of 5 are

$$5 \times 0 = 0$$

$$5 \times 1 = 5$$

$$5 \times 2 = 10$$

$$5 \times 3 = 15$$

$$5 \times 4 = 20$$

$$5 \times 5 = 25$$

etc.

The multiples of 6 are

$$6 \times 0 = 0$$

$$6 \times 1 = 6$$

$$6 \times 2 = 12$$

$$6 \times 3 = 18$$

$$6 \times 4 = 24$$

$$6 \times 5 = 30$$

etc.

Now, if you want to find a common denominator, simply list the multiples of the numbers in the denominators until a common multiple of both numbers is found. For example, the common denominator of $\frac{1}{5}$ and $\frac{1}{6}$ is found as follows:

$$0, 5, 10, 15, 20, 25, 30$$

$$0, 6, 12, 18, 24, 30$$

Since 30 is the smallest common multiple of 5 and 6, it is the lowest common denominator of $\frac{1}{5}$ and $\frac{1}{6}$.

When the denominators of the fractions are large, another method, called the **division method**, can be used. The procedure for finding the lowest common denominator of two or more fractions is as follows:

- *Step 1: Arrange the numbers in the denominators in a row.*
- *Step 2: Divide by the smallest number that divides evenly into two or more of the numbers.*
- *Step 3: Bring down to the next row all quotients and remaining numbers not used.*
- *Step 4: Continue dividing until no two of the denominators can be divided evenly by any number other than one.*
- *Step 5: Multiply all the divisors and the remaining numbers to get the lowest common denominator.*

EXAMPLE

Find the lowest common denominator of the fractions $\dfrac{3}{16}$, $\dfrac{1}{8}$, and $\dfrac{13}{24}$.

SOLUTION

Arrange the numbers from the denominators in a row and start dividing, as shown:

$$
\begin{array}{r|ccc}
2) & 16 & 8 & 24 \\ \hline
2) & 8 & 4 & 12 \\ \hline
2) & 4 & 2 & 6 \\ \hline
 & 2 & 1 & 3
\end{array}
$$

After dividing by 2 three times, there are no two numbers in the last row that can be divided by any number other than one, so you stop. Multiply the divisors and the numbers in the bottom row to get the lowest common denominator. $2 \times 2 \times 2 \times 2 \times 1 \times 3 = 48$. Hence, the lowest common denominator of the fractions $\dfrac{3}{16}$, $\dfrac{1}{8}$, and $\dfrac{13}{24}$ is 48.

EXAMPLE

Find the lowest common denominator of $\dfrac{1}{20}$, $\dfrac{7}{15}$, and $\dfrac{11}{30}$.

SOLUTION

$$
\begin{array}{r|ccc}
2) & 20 & 15 & 30 \\ \hline
3) & 10 & 15 & 15 \\ \hline
5) & 10 & 5 & 5 \\ \hline
 & 2 & 1 & 1
\end{array}
$$

Hence, the LCD is $2 \times 3 \times 5 \times 2 \times 1 \times 1 = 60$.

Practice

Find the LCD of each:

1. $\dfrac{2}{5}$ and $\dfrac{7}{8}$

2. $\dfrac{7}{18}$ and $\dfrac{15}{24}$

3. $\dfrac{13}{20}$ and $\dfrac{5}{36}$

4. $\dfrac{5}{28}$, $\dfrac{9}{16}$, and $\dfrac{7}{10}$

5. $\dfrac{5}{6}$, $\dfrac{7}{15}$, and $\dfrac{2}{25}$

Answers

1. 40
2. 72
3. 180
4. 560
5. 150

Addition of Fractions

In order to add two or more fractions, the fractions must have the same denominator.

To add two or more fractions:

- *Step 1: Find the LCD of the fractions.*
- *Step 2: Change the fractions to equivalent fractions in higher terms with the LCD.*
- *Step 3: Add the numerators of the fractions and write them over the LCD.*
- *Step 4: Reduce or simplify the answer if possible.*

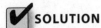 **EXAMPLE**

Add $\dfrac{5}{12} + \dfrac{11}{12}$.

✔ **SOLUTION**

Since both fractions have the same denominators, add the numerators and reduce the fractions.

$$\frac{5}{12} + \frac{11}{12} = \frac{16}{12} = 1\frac{4}{12} = 1\frac{1}{3}$$

 EXAMPLE

Add $\dfrac{7}{8} + \dfrac{3}{20}$.

✔ **SOLUTION**

Step 1: Find the LCD. It is 40.

Step 2: Change the fractions to higher terms:

$$\frac{7}{8} = \frac{35}{40} \qquad \frac{3}{20} = \frac{6}{40}$$

Step 3: Add the numerators:

$$\frac{35}{40} + \frac{6}{40} = \frac{41}{40}$$

Step 4: Simplify the answers:

$$\frac{41}{40} = 1\frac{1}{40}$$

EXAMPLE

Add $\dfrac{1}{4} + \dfrac{2}{3} + \dfrac{3}{8}$.

✔ SOLUTION

Addition can be done vertically as shown:

$$\frac{1}{4} = \frac{6}{24}$$

$$\frac{2}{3} = \frac{16}{24}$$

$$+\ \frac{3}{8} = \frac{9}{24}$$

$$\frac{31}{24} = 1\frac{7}{24}$$

Still Struggling

Never add the denominators of fractions.

Practice

Add each of the following fractions:

1. $\dfrac{7}{13} + \dfrac{3}{13}$

2. $\dfrac{15}{32} + \dfrac{3}{16}$

3. $\dfrac{9}{16} + \dfrac{5}{8} + \dfrac{1}{3}$

4. $\dfrac{1}{4} + \dfrac{5}{6} + \dfrac{7}{12}$

5. $\dfrac{5}{12} + \dfrac{11}{16} + \dfrac{7}{20}$

Answers

1. $\dfrac{10}{13}$

2. $\dfrac{21}{32}$

3. $1\dfrac{25}{48}$

4. $1\dfrac{2}{3}$

5. $1\dfrac{109}{240}$

Addition of Mixed Numbers

To add mixed numbers, first add the fractions and then add the whole numbers; simplify the answer, if necessary.

EXAMPLE

Add $6\dfrac{1}{8}+2\dfrac{2}{3}+5\dfrac{5}{6}$.

SOLUTION

$$6\dfrac{1}{8}=6\dfrac{3}{24}$$

$$2\dfrac{2}{3}=2\dfrac{16}{24}$$

$$+\ 5\dfrac{5}{6}=5\dfrac{20}{24}$$

$$13\dfrac{39}{24}=13+1\dfrac{15}{24}=14\dfrac{15}{24}=14\dfrac{5}{8}$$

EXAMPLE

Add $6\dfrac{3}{8}+11\dfrac{1}{5}+9\dfrac{2}{3}$.

SOLUTION

$$6\dfrac{3}{8}=6\dfrac{45}{120}$$

$$11\dfrac{1}{5}=11\dfrac{24}{120}$$

$$+\ 9\dfrac{2}{3}=9\dfrac{80}{120}$$

$$26\dfrac{149}{120}=26+1\dfrac{29}{120}=27\dfrac{29}{120}$$

Practice
Add:

1. $7\dfrac{1}{3} + 2\dfrac{7}{9}$

2. $9\dfrac{5}{16} + 14\dfrac{1}{6}$

3. $15\dfrac{7}{12} + 19\dfrac{5}{8}$

4. $7\dfrac{1}{6} + 3\dfrac{7}{8} + 8\dfrac{1}{4}$

5. $26\dfrac{9}{16} + 8\dfrac{7}{12} + 2\dfrac{3}{8}$

Answers

1. $10\dfrac{1}{9}$

2. $23\dfrac{23}{48}$

3. $35\dfrac{5}{24}$

4. $19\dfrac{7}{24}$

5. $37\dfrac{25}{48}$

Subtraction of Fractions

The rule for subtracting fractions is similar to the rule for adding fractions.
To subtract two fractions:

- *Step 1: Find the LCD of the fractions.*
- *Step 2: Change the fractions to equivalent fractions in higher terms with the LCD.*
- *Step 3: Subtract the numerators of the fractions and write the result over the LCD.*
- *Step 4: Reduce or simplify the answer if possible.*

EXAMPLE

Subtract $\dfrac{5}{6} - \dfrac{1}{4}$.

SOLUTION

$$\dfrac{5}{6} = \dfrac{10}{12}$$

$$-\dfrac{1}{4} = \dfrac{3}{12}$$

$$\dfrac{7}{12}$$

EXAMPLE

Subtract $\dfrac{9}{10} - \dfrac{5}{8}$.

SOLUTION

$$\dfrac{9}{10} = \dfrac{36}{40}$$

$$-\dfrac{5}{8} = \dfrac{25}{40}$$

$$\dfrac{11}{40}$$

Practice

Subtract:

1. $\dfrac{4}{5} - \dfrac{1}{5}$

2. $\dfrac{3}{4} - \dfrac{5}{12}$

3. $\dfrac{2}{3} - \dfrac{3}{16}$

4. $\dfrac{15}{16} - \dfrac{7}{20}$

5. $\dfrac{5}{6} - \dfrac{3}{8}$

Answers

1. $\dfrac{3}{5}$

2. $\dfrac{1}{3}$

3. $\dfrac{23}{48}$

4. $\dfrac{47}{80}$

5. $\dfrac{11}{24}$

Subtraction of Mixed Numbers

Subtraction of mixed numbers is a little more complicated than addition of mixed numbers since it is sometimes necessary to **borrow**.

To subtract two mixed numbers:

- *Step 1: Find the LCD of the fractions.*
- *Step 2: Change the fractions to higher terms with the LCD.*
- *Step 3: Subtract the fractions, borrowing if necessary.*
- *Step 4: Subtract the whole numbers.*
- *Step 5: Reduce or simplify the answer if necessary.*

When the fraction in the subtrahend is smaller than the fraction in the minuend, borrowing is not necessary.

EXAMPLE

Subtract $8\dfrac{2}{3} - 3\dfrac{1}{5}$.

SOLUTION

$$8\dfrac{2}{3} = 8\dfrac{10}{15}$$
$$-\;3\dfrac{1}{5} = 3\dfrac{3}{15}$$
$$\overline{\phantom{-\;3\dfrac{1}{5}=}\;5\dfrac{7}{15}}$$

Principles of Borrowing

When the fraction in the subtrahend is larger than the fraction in the minuend, it is necessary to borrow from the whole number.

When borrowing is necessary, take one (1) away from the whole number, change it to a fraction with the same numerator and denominator, and then add it to the fraction.

 EXAMPLE

Borrow 1 from $7\frac{1}{6}$.

 **SOLUTION**

$$7\frac{1}{6} = 6 + 1 + \frac{1}{6} \qquad \text{borrow 1 from 7}$$

$$= 6 + \frac{6}{6} + \frac{1}{6} \qquad \text{change 1 to } \frac{6}{6}$$

$$= 6\frac{7}{6} \qquad \text{add } \frac{6}{6} + \frac{1}{6}$$

 EXAMPLE

Borrow 1 from $5\frac{3}{8}$.

 SOLUTION

$$5\frac{3}{8} = 4 + 1 + \frac{3}{8}$$

$$= 4 + \frac{8}{8} + \frac{3}{8}$$

$$= 4\frac{11}{8}$$

The next examples show how to use borrowing when subtracting mixed numbers.

EXAMPLE

Subtract $10\frac{1}{6} - 3\frac{4}{5}$.

SOLUTION

$$10\frac{1}{6} = 10\frac{5}{30} = 9\frac{35}{30}$$

$$-\ 3\frac{4}{5} = 3\frac{24}{30} = 3\frac{24}{30}$$

$$6\frac{11}{30}$$

EXAMPLE

Subtract $14\frac{1}{3} - 2\frac{8}{9}$.

SOLUTION

$$14\frac{1}{3} = 14\frac{3}{9} = 13\frac{12}{9}$$

$$-\ 2\frac{8}{9} = 2\frac{8}{9} = \ 2\frac{8}{9}$$

$$11\frac{4}{9}$$

EXAMPLE

Subtract $11 - 6\frac{5}{6}$.

SOLUTION

$$11 = 10\frac{6}{6}$$

$$-\ 6\frac{5}{6} = \ 6\frac{5}{6}$$

$$4\frac{1}{6}$$

Still Struggling

A shortcut method can be used in borrowing. Simply reduce the whole number by one and then add the numerator and denominator together to get the new numerator and use the same denominator. For example, to borrow 1 from the number $3\frac{5}{8}$, $3\frac{5}{8} = 2\frac{8+5}{8} = 2\frac{13}{8}$.

Practice

Subtract:

1. $15\frac{7}{8} - 3\frac{5}{8}$

2. $29\frac{3}{5} - 14\frac{1}{8}$

3. $33\frac{17}{20} - 22\frac{3}{10}$

4. $14\frac{27}{32} - 3\frac{3}{16}$

5. $9\frac{3}{4} - 5$

6. $16\frac{2}{3} - 14\frac{2}{3}$

7. $12\frac{3}{10} - 7\frac{5}{8}$

8. $28\frac{7}{12} - 11\frac{7}{8}$

9. $16\frac{2}{5} - 10\frac{11}{12}$

10. $31 - 17\frac{4}{5}$

Answers

1. $12\dfrac{1}{4}$

2. $15\dfrac{19}{40}$

3. $11\dfrac{11}{20}$

4. $11\dfrac{21}{32}$

5. $4\dfrac{3}{4}$

6. 2

7. $4\dfrac{27}{40}$

8. $16\dfrac{17}{24}$

9. $5\dfrac{29}{60}$

10. $13\dfrac{1}{5}$

Multiplication of Fractions

Multiplying fractions uses the principle of **cancellation.** Cancellation saves you from reducing the answer after multiplying.

To cancel, divide any numerator and any denominator by the largest number possible (i.e., the greatest common factor).

To multiply two or more fractions:

- *Step 1: Cancel if possible.*
- *Step 2: Multiply numerators.*
- *Step 3: Multiply denominators.*
- *Step 4: Simplify the answer if possible.*

EXAMPLE

Multiply $\dfrac{5}{6} \times \dfrac{18}{25}$.

SOLUTION

$$\frac{5}{6} \times \frac{18}{25} = \frac{\overset{1}{\cancel{5}}}{\cancel{6}} \times \frac{\overset{3}{\cancel{18}}}{\cancel{25}}$$

$$\phantom{\frac{5}{6} \times \frac{18}{25}} \underset{1}{} \quad \underset{5}{}$$

$$= \frac{1 \times 3}{1 \times 5}$$

$$= \frac{3}{5}$$

EXAMPLE

Multiply $\dfrac{3}{4} \times \dfrac{8}{15} \times \dfrac{2}{3}$.

SOLUTION

$$\frac{3}{4} \times \frac{8}{15} \times \frac{2}{3} = \frac{\overset{1}{\cancel{3}}}{\cancel{4}} \times \frac{\overset{2}{\cancel{8}}}{15} \times \frac{2}{\cancel{3}}$$

$$= \frac{1 \times 2 \times 2}{1 \times 15 \times 1}$$

$$= \frac{4}{15}$$

MATH NOTE *Many times there is more than one way to cancel.*

Practice
Multiply:

1. $\dfrac{15}{16} \times \dfrac{12}{25}$

2. $\dfrac{7}{8} \times \dfrac{4}{35}$

3. $\dfrac{8}{9} \times \dfrac{3}{4}$

4. $\dfrac{7}{12} \times \dfrac{2}{21} \times \dfrac{3}{8}$

5. $\dfrac{3}{4} \times \dfrac{5}{6} \times \dfrac{8}{15}$

Answers

1. $\dfrac{9}{20}$

2. $\dfrac{1}{10}$

3. $\dfrac{2}{3}$

4. $\dfrac{1}{48}$

5. $\dfrac{1}{3}$

Multiplication of Mixed Numbers

Multiplying mixed numbers is similar to multiplying fractions.

To multiply two or more mixed numbers, first change the mixed numbers to improper fractions and then follow the steps for multiplying fractions.

EXAMPLE

Multiply $1\dfrac{5}{7} \times 2\dfrac{5}{8}$.

**SOLUTION**

$$1\dfrac{5}{7} \times 2\dfrac{5}{8} = \dfrac{12}{7} \times \dfrac{21}{8}$$

$$= \dfrac{\overset{3}{\cancel{12}}}{\underset{1}{\cancel{7}}} \times \dfrac{\overset{3}{\cancel{21}}}{\underset{2}{\cancel{8}}}$$

$$= \dfrac{3 \times 3}{1 \times 2}$$

$$= \dfrac{9}{2} = 4\dfrac{1}{2}$$

MATH NOTE   *When a whole number is used in a multiplication problem, make it into a fraction with a denominator of 1.*

EXAMPLE

Multiply $15 \times 2\frac{2}{5}$.

SOLUTION

$$15 \times 2\frac{2}{5} = \frac{15}{1} \times \frac{12}{5}$$

$$= \frac{\overset{3}{\cancel{15}}}{1} \times \frac{12}{\underset{1}{\cancel{5}}}$$

$$= \frac{3 \times 12}{1 \times 1}$$

$$= \frac{36}{1} = 36$$

EXAMPLE

Multiply $2\frac{1}{4} \times 6\frac{1}{3} \times 3\frac{1}{5}$.

SOLUTION

$$2\frac{1}{4} \times 6\frac{1}{3} \times 3\frac{1}{5} = \frac{9}{4} \times \frac{19}{3} \times \frac{16}{5}$$

$$= \frac{\overset{3}{\cancel{9}}}{\underset{1}{\cancel{4}}} \times \frac{19}{\underset{1}{\cancel{3}}} \times \frac{\overset{4}{\cancel{16}}}{5}$$

$$= \frac{3 \times 19 \times 4}{1 \times 1 \times 5}$$

$$= \frac{228}{5}$$

$$= 45\frac{3}{5}$$

Practice

Multiply:

1. $6\dfrac{1}{5} \times 3\dfrac{1}{3}$

2. $4\dfrac{1}{8} \times 6\dfrac{3}{11}$

3. $16 \times 2\dfrac{3}{10}$

4. $1\dfrac{3}{4} \times 2\dfrac{3}{8} \times 7\dfrac{5}{6}$

5. $3\dfrac{2}{3} \times 5\dfrac{1}{6} \times 4\dfrac{3}{8}$

Answers

1. $20\dfrac{2}{3}$

2. $25\dfrac{7}{8}$

3. $36\dfrac{4}{5}$

4. $32\dfrac{107}{192}$

5. $82\dfrac{127}{144}$

Division of Fractions

When dividing fractions, it is necessary to use the **reciprocal** of a fraction. To find the reciprocal of a fraction, interchange the numerator and the denominator. For example, the reciprocal of $\dfrac{3}{4}$ is $\dfrac{4}{3}$. The reciprocal of $\dfrac{5}{9}$ is $\dfrac{9}{5}$. The reciprocal of 16 is $\dfrac{1}{16}$. Finding the reciprocal of a fraction is also called **inverting**.

To divide two fractions, invert the fraction after the division sign and multiply.

EXAMPLE

Divide $\dfrac{2}{3} \div \dfrac{4}{9}$.

SOLUTION

$$\frac{2}{3} \div \frac{4}{9} = \frac{2}{3} \times \frac{9}{4} \qquad \text{invert } \frac{4}{9} \text{ and multiply}$$

$$= \frac{\overset{1}{\cancel{2}}}{\underset{1}{\cancel{3}}} \times \frac{\overset{3}{\cancel{9}}}{\underset{2}{\cancel{4}}}$$

$$= \frac{3}{2} = 1\frac{1}{2}$$

EXAMPLE

Divide $\dfrac{4}{15} \div \dfrac{2}{3}$.

SOLUTION

$$\frac{4}{15} \div \frac{2}{3} = \frac{4}{15} \times \frac{3}{2}$$

$$= \frac{\overset{2}{\cancel{4}}}{\underset{5}{\cancel{15}}} \times \frac{\overset{1}{\cancel{3}}}{\underset{1}{\cancel{2}}}$$

$$= \frac{2}{5}$$

MATH NOTE *When a number (except 0) is multiplied by its reciprocal, the answer will always be 1. For example,* $\dfrac{5}{6} \times \dfrac{6}{5} = \dfrac{\overset{1}{\cancel{5}}}{\underset{1}{\cancel{6}}} \times \dfrac{\overset{1}{\cancel{6}}}{\underset{1}{\cancel{5}}} = \dfrac{1}{1} = 1.$

Practice

Divide:

1. $\dfrac{15}{16} \div \dfrac{3}{8}$

2. $\dfrac{3}{10} \div \dfrac{9}{40}$

3. $\dfrac{3}{4} \div \dfrac{5}{6}$

4. $\dfrac{5}{8} \div \dfrac{5}{16}$

5. $\dfrac{5}{9} \div \dfrac{2}{3}$

Answers

1. $2\dfrac{1}{2}$

2. $1\dfrac{1}{3}$

3. $\dfrac{9}{10}$

4. 2

5. $\dfrac{5}{6}$

Division of Mixed Numbers

To divide mixed numbers, change them into improper fractions, invert the fraction after the division sign, and multiply.

 EXAMPLE

Divide $4\dfrac{4}{5} \div 3\dfrac{1}{3}$.

✔ SOLUTION

$$4\frac{4}{5} \div 3\frac{1}{3} = \frac{24}{5} \div \frac{10}{3}$$

$$= \frac{24}{5} \times \frac{3}{10}$$

$$= \frac{\overset{12}{\cancel{24}}}{5} \times \frac{3}{\underset{5}{\cancel{10}}}$$

$$= \frac{36}{25} = 1\frac{11}{25}$$

EXAMPLE

Divide $3\frac{3}{4} \div 2\frac{1}{2}$.

✔ SOLUTION

$$3\frac{3}{4} \div 2\frac{1}{2} = \frac{15}{4} \div \frac{5}{2}$$

$$= \frac{15}{4} \times \frac{2}{5}$$

$$= \frac{\overset{3}{\cancel{15}}}{\underset{2}{\cancel{4}}} \times \frac{\overset{1}{\cancel{2}}}{\underset{1}{\cancel{5}}}$$

$$= \frac{3}{2} = 1\frac{1}{2}$$

Practice

Divide:

1. $5\frac{3}{5} \div 1\frac{3}{7}$

2. $3\frac{1}{6} \div 3\frac{3}{4}$

3. $1\dfrac{7}{9} \div 3\dfrac{2}{3}$

4. $5\dfrac{5}{8} \div 6\dfrac{3}{4}$

5. $9\dfrac{1}{6} \div 8\dfrac{4}{5}$

Answers

1. $3\dfrac{23}{25}$

2. $\dfrac{38}{45}$

3. $\dfrac{16}{33}$

4. $\dfrac{5}{6}$

5. $1\dfrac{1}{24}$

Word Problems

Word problems involving fractions are done using the same procedure as those involving whole numbers.

EXAMPLE

A carpenter needs to cut three pieces of wood that measure $6\dfrac{1}{2}$ inches, $5\dfrac{3}{4}$ inches, and $3\dfrac{5}{8}$ inches. How long of a board is needed to cut all three pieces from it?

SOLUTION

It is necessary to find the total length, so we add the measurements:

$$6\frac{1}{2} = 6\frac{4}{8}$$

$$5\frac{3}{4} = 5\frac{6}{8}$$

$$+ \ 3\frac{5}{8} = 3\frac{5}{8}$$

$$= 14\frac{15}{8} = 15\frac{7}{8} \text{ inches}$$

The carpenter will need a piece at least $15\frac{7}{8}$ inches long. (Do not consider waste.)

EXAMPLE

Lemont worked $6\frac{3}{5}$ hours on Thursday and $4\frac{3}{4}$ hours on Saturday. How much longer did he work on Thursday than he did on Saturday?

 SOLUTION

In this case, we are looking for a difference, so we subtract:

$$6\frac{3}{5} = 6\frac{12}{20} = 5\frac{32}{20}$$

$$- \ 4\frac{3}{4} = 4\frac{15}{20} = 4\frac{15}{20}$$

$$= 1\frac{17}{20}$$

He worked $1\frac{17}{20}$ of an hour longer on Thursday.

EXAMPLE

It was found that $\frac{3}{5}$ of all waste products generated consists of paper products. How many tons of waste products were generated from paper if a municipality generated 8,500 tons of waste materials in a week?

 **SOLUTION**

In this case, it is necessary to multiply.

$$8500 \times \frac{3}{5} = \frac{8500}{1} \times \frac{3}{5}$$

$$= \frac{\overset{1700}{\cancel{8500}}}{1} \times \frac{3}{\cancel{5}}$$
$$\qquad\qquad 1$$

$$= 5{,}100 \text{ tons}$$

The waste products consisted of 5,100 tons of paper.

 EXAMPLE

How many cubic feet of water are contained in a 300 gallon tank if 1 cubic foot of water is about $7\frac{1}{2}$ gallons?

 **SOLUTION**

Divide:

$$300 \div 7\frac{1}{2} = \frac{300}{1} \div \frac{15}{2}$$

$$= \frac{\overset{20}{\cancel{300}}}{1} \times \frac{2}{\cancel{15}}$$
$$\qquad\qquad 1$$

$$= 40 \text{ cubic feet}$$

The tank will hold about 40 cubic feet of water.

Practice

1. A company uses $\frac{3}{5}$ of its budget for advertising. If $\frac{1}{6}$ of that amount was used for newspaper ads, what fractional part of its budget is used for newspaper advertising?

2. If a bolt $2\frac{1}{2}$ inches long is placed through a piece of wood that is $1\frac{5}{8}$ inches thick, how much of the bolt will be extending out of the wood?

3. A recipe for 4 servings calls for $3\frac{3}{4}$ cups of flour. If the chef wanted to make a single serving, how much flour would she use?

4. A plumber needs to cut 4 pieces of pipe measuring $6\frac{3}{5}$ inches, $2\frac{7}{8}$ inches, $3\frac{3}{4}$ inches, and $5\frac{1}{2}$ inches, respectively. How much pipe is needed to cut these pieces? (Ignore waste.)

5. A \$16,000 estate was to be divided among 3 people. The first person received $\frac{2}{5}$ of the money, the second person received $\frac{1}{4}$ of the money, and the third person received the rest. How much did each person receive?

Answers

1. $\frac{1}{10}$

2. $\frac{7}{8}$ inch

3. $\frac{15}{16}$ of a cup

4. $18\frac{29}{40}$ inches

5. \$6,400, \$4,000, and \$5,600

Comparing Fractions

To compare two or more fractions:

- *Step 1: Find the LCD of the fractions.*
- *Step 2: Change the fractions to higher terms using the LCD.*
- *Step 3: Compare the numerators.*

 EXAMPLE

Which is larger $\frac{3}{5}$ or $\frac{4}{7}$?

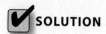 SOLUTION _____

$$\frac{3}{5} = \frac{21}{35} \quad \text{and} \quad \frac{4}{7} = \frac{20}{35}$$

Since 21 is larger than 20, the fraction $\frac{3}{5}$ is larger than $\frac{4}{7}$.

Practice
Which is larger:

1. $\frac{5}{6}$ or $\frac{7}{9}$?

2. $\frac{11}{16}$ or $\frac{3}{4}$?

3. $\frac{8}{9}$ or $\frac{11}{14}$?

4. $\frac{9}{32}$ or $\frac{5}{8}$?

5. $\frac{2}{3}$ or $\frac{7}{9}$?

Answers

1. $\frac{5}{6}$

2. $\frac{3}{4}$

3. $\frac{8}{9}$

4. $\frac{5}{8}$

5. $\frac{7}{9}$

Operations with Positive and Negative Fractions

Every fraction has three signs: the sign of the number in the numerator, the sign of the number in the denominator, and the sign in front of the fraction (called the sign of the fraction).

When a fraction is negative, such as $-\dfrac{2}{5}$, the negative sign can be placed in the numerator or denominator. For example,

$$-\frac{2}{5} = \frac{-2}{5} = \frac{2}{-5}$$

 ## Still Struggling

When performing operations with fractions that are negative, keep the negative sign with the numerator of the fraction and perform the operations.

▢ **EXAMPLE**

Add $-\dfrac{7}{8} + \dfrac{1}{3}$.

✔ **SOLUTION**

$$-\frac{7}{8} + \frac{1}{3} = \frac{-7}{8} + \frac{1}{3}$$

$$= \frac{-21}{24} + \frac{8}{24}$$

$$= \frac{-21 + 8}{24}$$

$$= \frac{-13}{24} = -\frac{13}{24}$$

▢ **EXAMPLE**

Subtract $-\dfrac{7}{12} - \dfrac{5}{8}$.

✔ **SOLUTION**

$$-\frac{7}{12} - \frac{5}{8} = \frac{-7}{12} - \frac{5}{8} = \frac{-14}{24} - \frac{15}{24}$$

$$= \frac{-14 - 15}{24}$$

$$= \frac{-29}{24} = -1\frac{5}{24}$$

EXAMPLE

Multiply $\dfrac{7}{8} \times \left(-\dfrac{16}{21}\right)$.

SOLUTION

$$\dfrac{7}{8} \times \left(-\dfrac{16}{21}\right) = \dfrac{7}{8} \times \dfrac{-16}{21}$$

$$= \dfrac{\overset{1}{7}}{\underset{1}{8}} \times \dfrac{\overset{-2}{-16}}{\underset{3}{21}}$$

$$= \dfrac{-2}{3} = -\dfrac{2}{3}$$

EXAMPLE

Divide $\dfrac{3}{4} \div \left(-\dfrac{3}{8}\right)$.

SOLUTION

$$\dfrac{3}{4} \div \left(-\dfrac{3}{8}\right) = \dfrac{3}{4} \div \dfrac{-3}{8}$$

$$= \dfrac{\overset{1}{3}}{\underset{1}{4}} \times \dfrac{\overset{2}{8}}{\underset{-1}{-3}}$$

$$= \dfrac{2}{-1} = -2$$

Practice
Perform the indicated operation:

1. $\dfrac{5}{6} + \left(-\dfrac{1}{2}\right) - \dfrac{3}{8}$

2. $\dfrac{7}{10} - \left(-\dfrac{2}{5}\right)$

3. $\dfrac{3}{8} \times \left(-\dfrac{2}{5}\right) \times \left(-\dfrac{1}{6}\right)$

4. $\dfrac{17}{24} \div \left(-\dfrac{3}{4}\right)$

5. $-\dfrac{2}{3} \times \dfrac{5}{9} - \dfrac{1}{6}$

Answers

1. $-\dfrac{1}{24}$

2. $1\dfrac{1}{10}$

3. $\dfrac{1}{40}$

4. $-\dfrac{17}{18}$

5. $-\dfrac{29}{54}$

In this chapter, you learned how to find a common denominator of two or more fractions and how to add or subtract fractions and mixed numbers. In addition, multiplication and division of fractions and mixed numbers were explained. Finally, operations with positive and negative fractions were shown.

QUIZ

1. Add $\frac{5}{9} + \frac{2}{3}$.

   $\frac{2}{3} \times \frac{8}{\times 3} = \frac{6}{9} + \frac{5}{9} = \frac{11}{9} \;\Big|\; \frac{2}{9}$

   A. $\frac{7}{12}$

   B. $1\frac{2}{9}$ (B)

   C. $\frac{9}{11}$

   D. $2\frac{1}{9}$

2. Subtract $\frac{7}{10} - \frac{1}{6}$.

   $10: 20, 30$
   $6: 12, 18, 24, 30$

   A. $\frac{8}{15}$

   $\frac{7}{10} \times \frac{3}{3} = \frac{21}{30}$ $\frac{1}{6} \times \frac{5}{5} = \frac{5}{30}$ $2\overline{)30}^{15} \; -\frac{24}{1}$

   B. $\frac{1}{9}$

   C. $\frac{1}{2}$ (A) $\frac{21}{30} - \frac{5}{30} = \frac{16}{30} \div \frac{2}{2} = \frac{8}{15}$

   D. $1\frac{1}{2}$

3. Multiply $\frac{5}{8} \times \frac{4}{15}$.

   $8: 16, 24, 32, 40, 48, 56, 64, 72, 80, 88$
   $15: 30, 45, 60, 75$

   A. $\frac{9}{23}$

   B. $\frac{21}{40}$ $\frac{5 \times 15}{8 \times 16} = \frac{80}{120}$ $\frac{4 \times 8}{15 \times 8} = \frac{32}{120}$

   C. $\frac{9}{120}$ $\frac{256 \div 2}{120 \div 2} = \frac{128 \div 4}{60 \div 4} = \frac{32}{15}$

   D. $\frac{1}{6}$

4. Divide $\frac{5}{9} \div \frac{5}{6}$.

   A. $\frac{1}{3}$ $9 \times 2 = 18$
   $6 \times 3 = 18$

   B. $\frac{25}{54}$ $10 \div 15 =$

   C. $\frac{2}{3}$ $\frac{5 \times 2}{9 \times 2} = \frac{10}{18}$ $\frac{5 \times 3}{6 \times 3} = \frac{15}{18}$

   D. $\frac{1}{15}$

   $\frac{10}{18} \times \frac{15}{18} = \frac{150 \div 3}{18 \div 3} = \frac{50}{6} = \frac{25 \div}{3}$

5. Add $3\dfrac{7}{8} + 2\dfrac{2}{3} + 4\dfrac{9}{10}$.

A. $11\dfrac{53}{120}$

B. $15\dfrac{19}{60}$

C. $9\dfrac{6}{7}$

D. $10\dfrac{11}{24}$

6. Subtract $19\dfrac{7}{12} - 6\dfrac{8}{9}$.

A. $13\dfrac{1}{3}$

B. $12\dfrac{3}{4}$

C. $13\dfrac{11}{36}$

D. $12\dfrac{25}{36}$

7. Multiply $10\dfrac{2}{3} \times 5\dfrac{1}{6}$.

A. $50\dfrac{1}{9}$

B. $55\dfrac{1}{9}$

C. $15\dfrac{5}{6}$

D. $2\dfrac{2}{31}$

8. Divide $8\dfrac{1}{3} \div 12\dfrac{3}{5}$.

A. $\dfrac{125}{189}$

B. $96\dfrac{1}{5}$

C. 105

D. $\dfrac{126}{163}$

9. Add $6\frac{3}{4} + 2\frac{1}{2} + \frac{5}{6}$.

   A. $8\frac{2}{3}$

   B. $9\frac{5}{12}$

   C. $9\frac{7}{12}$

   D. $10\frac{1}{12}$

10. Simplify $1\frac{3}{5} + \frac{3}{4} \times \frac{2}{3}$.

    A. $1\frac{17}{30}$

    B. $3\frac{1}{60}$

    C. $2\frac{1}{10}$

    D. $\frac{4}{5}$

11. Last month Madison weighed $129\frac{2}{3}$ pounds. This month she weighed $125\frac{1}{2}$ pounds. How much weight did she lose?

    A. $3\frac{1}{8}$ pounds

    B. $5\frac{1}{3}$ pounds

    C. $3\frac{5}{6}$ pounds

    D. $4\frac{1}{6}$ pounds

12. A person bought $6\frac{3}{4}$ ounces of decaf coffee, $5\frac{1}{2}$ ounces of caramel coffee, and $10\frac{2}{3}$ ounces of mocha coffee. How many ounces did the person buy altogether?

    A. $22\frac{3}{8}$ ounces

    B. $21\frac{7}{8}$ ounces

C. $22\frac{11}{12}$ ounces

D. $22\frac{1}{4}$ ounces

13. How many $\frac{3}{4}$ pound bags can be filled with sugar from a container which holds $32\frac{1}{4}$ pounds?

   A. 43
   B. 42
   C. 45
   D. $44\frac{1}{3}$

14. Find the cost of $3\frac{1}{2}$ pounds of bananas at 60 cents per pound.

   A. 125 cents
   B. 210 cents
   C. 155 cents
   D. 135 cents

15. Which fraction is the smallest: $\frac{2}{3}, \frac{7}{16}, \frac{3}{4}$, or $\frac{3}{8}$?

   A. $\frac{2}{3}$

   B. $\frac{7}{16}$

   C. $\frac{3}{4}$

   D. $\frac{3}{8}$

16. Add $\frac{5}{9} + \left(-\frac{3}{4}\right) + \frac{5}{6}$.

   A. $\frac{13}{36}$

   B. $1\frac{5}{36}$

   C. $\frac{23}{36}$

   D. $-\frac{17}{36}$

17. **Subtract** $-6\frac{1}{4} - 2\frac{2}{3}$.

    A. $4\frac{5}{12}$

    B. $-8\frac{11}{12}$

    C. $-4\frac{11}{12}$

    D. $8\frac{7}{12}$

18. **Multiply** $\frac{3}{10} \times \left(-\frac{2}{3}\right) \times \left(-\frac{5}{6}\right)$.

    A. $\frac{1}{6}$

    B. $\frac{5}{6}$

    C. $-\frac{1}{6}$

    D. $-\frac{5}{6}$

19. **Divide** $\frac{5}{8} \div \left(-2\frac{1}{4}\right)$.

    A. $-10\frac{1}{6}$

    B. $-11\frac{17}{32}$

    C. $-10\frac{1}{32}$

    D. $-\frac{5}{18}$

20. **Perform the indicated operations:** $-\frac{5}{6} + \left(-\frac{3}{4}\right) \times 2\frac{1}{8}$.

    A. $2\frac{3}{4}$

    B. $2\frac{19}{20}$

    C. $-2\frac{41}{96}$

    D. $2\frac{47}{96}$

chapter 5

Decimals

This chapter explains the basic operations using decimals. Decimals are used when word problems involve money. For example, 5 dollars and 95 cents is written as $5.95. Decimals are also used quite often in chemistry and physics problems. For example, the acceleration of gravity on Earth at sea level is 9.8 meters per second squared. Unit conversions involve decimals. For example, 1 inch equals approximately 2.54 centimeters. Decimals are also used to solve percent problems.

CHAPTER OBJECTIVES

In this chapter, you will learn how to

- Read decimals
- Round decimals
- Add, subtract, multiply, and divide decimals
- Compare decimals
- Solve word problems using decimals
- Change fractions to decimals
- Change decimals to fractions
- Perform operations with fractions and decimals
- Perform operations with positive and negative decimals

Naming Decimals

As with whole numbers, each digit of a decimal has a **place value**. The place value names are shown in Fig. 5-1.

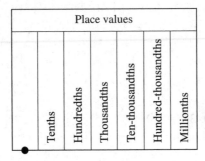

Place values					
Tenths	Hundredths	Thousandths	Ten-thousandths	Hundred-thousandths	Millionths

FIGURE 5-1

For example, in the number 0.6235, the 2 is in the hundredths place. The 5 is in the ten-thousandths place.

When naming a decimal, read the number from left to right as you would read a whole number, and then use the place value name for the last digit of the number.

EXAMPLE

Name 0.3462.

SOLUTION

First write in words 3,462, and then write the place value of the 2 after it: three thousand four hundred sixty-two ten-thousandths.

EXAMPLE

Name 0.000015.

SOLUTION

Fifteen millionths.

When there are digits to the left and to the right of the decimal point, the decimal part is written using the word "and."

EXAMPLE

Name 93.731.

SOLUTION

Ninety-three and seven hundred thirty-one thousandths.

Practice

Name each number:

1. 0.666 6
2. 0.31472 thirty one thousands
3. 0.0043
4. 163.5
5. 18.72

Answers

1. Six hundred sixty-six thousandths
2. Thirty-one thousand four hundred seventy-two hundred-thousandths
3. Forty-three ten-thousandths
4. One hundred sixty-three and five tenths
5. Eighteen and seventy-two hundredths

Rounding Decimals

Decimals are rounded the same way whole numbers are rounded.

To round a decimal to a specific place value, first locate that place value digit in the number. If the digit to the right is 0, 1, 2, 3, or 4, the place value digit remains the same. If the digit to the right of the specific value digit is 5, 6, 7, 8, or 9, add one to the specific place value digit. All digits to the right of the place value digit are dropped.

EXAMPLE

Round 0.65376 to the nearest ten-thousandth.

SOLUTION

We are rounding to the ten-thousandths place, which is the digit 7. Since the digit to the right of the 7 is 6, raise the 7 to an 8 and drop all digits to the right of the 8. Hence, the answer is 0.6538.

EXAMPLE

Round 56.7329 to the nearest hundredth.

SOLUTION

The digit in the hundredths place is 3 and since the digit to the right of 3 is 2, the 3 remains the same. Hence, the answer is 56.73.

Practice

1. Round 0.4253 to the nearest thousandth. *0.4250*
2. Round 0.63275 to the nearest tenth. *0.60 000*
3. Round 59.0468 to the nearest hundredth. *59,0 500*
4. Round 0.3999 to the nearest thousandth. *0.39 100*
5. Round 4.976 to the nearest one. *5.*

Answers

1. 0.425
2. 0.6
3. 59.05
4. 0.4
5. 5

Still Struggling

Zeros can be affixed to the end of a decimal on the right side of the decimal point. For example, 0.62 can be written as 0.620 or 0.6200. Likewise, the zeros can be dropped if they are at the end of a decimal on the right side of the decimal point. For example, 0.3750 can be written as 0.375.

Addition of Decimals

In order to add two or more decimals, write the numbers in a column placing the decimal point of each number in a vertical line. Add the numbers and place the decimal point in the sum directly under the other decimal points above.

EXAMPLE

Add 31.93 + 257.806 + 7.7.

SOLUTION

```
   31.930      Zeros are used to keep the columns straight.
  257.806
+   7.700
  297.436
```

EXAMPLE

Add 0.06 + 22.638 + 847.2.

SOLUTION

```
    0.060
   22.638
+ 847.200
  869.898
```

Practice

Add:

1. 7.3 + 9.621 + 0.75
2. 143.78 + 0.117 + 27.9
3. 15 + 235.01 + 62.6 + 46.22
4. 0.00004 + 97.2 + 17.006
5. 375.9 + 4.8617 + 2.5

Answers

1. 17.671
2. 171.797
3. 358.83
4. 114.20604
5. 383.2617

Subtraction of Decimals

Subtracting decimals is similar to adding decimals.

To subtract two decimals, write the decimals in a column placing the decimal points in a vertical line. Subtract the numbers and place the decimal point in the difference directly under the other decimal points.

 EXAMPLE

Subtract 517.328 – 16.18.

 SOLUTION

$$
\begin{array}{r}
517.328 \\
-\ \ 16.180 \\
\hline
501.148
\end{array}
$$
   Insert a zero to keep the columns straight.

 EXAMPLE

Subtract 19.7 – 3.875.

 SOLUTION

$$
\begin{array}{r}
19.700 \\
-\ \ 3.875 \\
\hline
15.825
\end{array}
$$
   Insert zeros to keep the columns straight.

Practice
Subtract:

1. 18.681 – 13.297
2. 0.76 – 0.6246
3. 519.8 – 135.84
4. 14.65193 – 12.281
5. 9.2 – 3.67

Answers

1. 5.384
2. 0.1354

3. 383.96

4. 2.37093

5. 5.53

Multiplication of Decimals

To multiply two decimals, multiply the two numbers disregarding the decimal point and then count the total number of digits after the decimal points in both numbers. Count the same number of places from the right in the product and place the decimal point there. If there are fewer digits in the product than places, insert as many zeros as needed.

EXAMPLE

Multiply 81.6 × 0.28.

SOLUTION

$$
\begin{array}{r}
81.6 \\
\times\ 0.28 \\
\hline
6528 \\
\underline{1632\ } \\
22.848
\end{array}
$$

Three decimal places are needed in the answer

EXAMPLE

Multiply 71.3 × 0.0005.

SOLUTION

$$
\begin{array}{r}
71.3 \\
\times\ 0.0005 \\
\hline
0.03565
\end{array}
$$

Since five places are needed in the answer, it is necessary to insert one zero in front of the product.

MATH NOTE *A shortcut for multiplying a decimal by 10, 100, 1000, etc., is to move the decimal point to the right as many places as the zeros in the multiplier. For example,*

5.62 × 10 = 56.2

5.62 × 100 = 562

5.62 × 1000 = 5620

etc.

Practice
Multiply:

1. 562.6 × 8.05

2. 63.2 × 0.7

3. 39.62 × 5.7

4. 0.006 × 0.02

5. 162.3 × 0.614

Answers

1. 4528.93

2. 44.24

3. 225.834

4. 0.00012

5. 99.6522

Division of Decimals

When dividing two decimals, it is important to find the correct location of the decimal point in the quotient. There are two cases.

Case 1. To divide a decimal by a whole number, divide as though both numbers were whole numbers and place the decimal point in the quotient directly above the decimal point in the dividend.

**EXAMPLE**
Divide 2900.2 by 34.

✔ SOLUTION

$$\begin{array}{r} 85.3 \\ 34\overline{)2900.2} \\ \underline{272} \\ 180 \\ \underline{170} \\ 102 \\ \underline{102} \\ 0 \end{array}$$

EXAMPLE

Divide 4.16 by 26.

✔ **SOLUTION**

$$\begin{array}{r} .16 \\ 26\overline{)4.16} \\ \underline{26} \\ 156 \\ \underline{156} \\ 0 \end{array}$$

Sometimes it is necessary to insert zeros in the quotient.

EXAMPLE

Divide 0.0042 by 6.

✔ **SOLUTION**

$$\begin{array}{r} .0007 \\ 6\overline{)0.0042} \\ \underline{42} \\ 0 \end{array}$$

Case 2. When the divisor contains a decimal point, move the point to the right of the last digit in the divisor. Then move the point to the right the same number of places in the dividend. Divide and place the point in the quotient directly above the point in the dividend.

EXAMPLE
Divide 5.664 by 0.16.

SOLUTION

$$0.16\overline{)5.664}$$

Move the point two places to the right as shown:

$$
\begin{array}{r}
35.4 \\
16\overline{)566.4} \\
\underline{48} \\
86 \\
\underline{80} \\
64 \\
\underline{64} \\
0
\end{array}
$$

Sometimes it is necessary to place zeros in the dividend.

EXAMPLE
Divide 24 by 0.625.

SOLUTION

$$0.625\overline{)24}$$

Move the point three places to the right after annexing three zeros.

$$
\begin{array}{r}
38.4 \\
625\overline{)24000.0} \\
\underline{1875} \\
5250 \\
\underline{5000} \\
2500 \\
\underline{2500} \\
0
\end{array}
$$

Practice
Divide:

1. $393.4 \div 7$
2. $17.655 \div 55$
3. $1.332 \div 3.33$
4. $4.96 \div 0.62$
5. $0.00057 \div 19$

Answers

1. 56.2
2. 0.321
3. 0.4
4. 8
5. 0.00003

Comparing Decimals

To compare two or more decimals, place the numbers in a vertical column with the decimal points aligned under each other. Add zeros to the end of the decimals so that they all have the same number of decimal places. Then compare the numbers, ignoring the decimal points.

 EXAMPLE

Which is larger, 0.651 or 0.27?

 **SOLUTION**

$$0.651 = 0.651 \rightarrow 651$$
$$0.27 = 0.270 \rightarrow 270$$

Since 651 is larger than 270, 0.651 is larger than 0.27.

EXAMPLE

Arrange the decimals 5.6, 0.8, 0.317, and 0.72 in order of size, smallest to largest.

 **SOLUTION**

5.600	5600
0.800	800
0.317	317
0.720	720

In order: 0.317, 0.72, 0.8, and 5.6.

Practice

1. Which is larger: 0.039 or 0.01?

2. Which is smaller: 0.24 or 0.657?

3. Arrange in order (smallest first): 0.55, 0.5, 0.555.

4. Arrange in order (largest first): 0.375, 0.3752, 0.37.

5. Arrange in order (largest first): 7.0, 0.07, 0.7, 0.007.

Answers

1. 0.039

2. 0.24

3. 0.5, 0.55, 0.555

4. 0.3752, 0.375, 0.37

5. 7.0, 0.7, 0.07, 0.007

Word Problems

To solve a word problem involving decimals, follow the procedure given in Chap. 1:

1. Read the problem carefully.

2. Identify what you are being asked to find.

3. Perform the correct operation or operations.

4. Check your answer or at least see if it is reasonable.

 **EXAMPLE**

Normal body temperature is 98.6°F. If a person has a fever and his temperature is 102.4°F, how much higher is his temperature than normal?

 SOLUTION

		Check:
	102.4	98.6
	− 98.6	+ 3.8
	3.8	102.4

Hence, his temperature is 3.8°F higher than normal.

 **EXAMPLE**

If an SUV averages 32.5 miles per gallon, how many gallons of gasoline will be needed for a trip that is a distance of 520 miles?

 **SOLUTION**

Divide 520 by 32.5:

$$32.5\overline{)520}$$

$$\begin{array}{r} 16 \\ 325\overline{)5200} \\ \underline{325} \\ 1950 \\ \underline{1950} \\ 0 \end{array}$$

Check:

$$\begin{array}{r} 32.5 \\ \times\ 16 \\ \hline 1950 \\ \underline{325} \\ 520.0 \end{array}$$

Hence, 16 gallons will be needed.

Practice

1. A gas range costs $1,795.20. If it is purchased with no money down and 12 monthly payments are made, find the monthly payment.

2. Cecelia ran a 100-yard dash in 15.8 seconds. After practicing for a month, she ran the same distance in 12.9 seconds. How many seconds did she improve by practicing?

3. The National Weather Service found that the amount of rain for a particular city over a 4-month period was 6.53 inches, 2.7 inches, 3 inches, and 4.8 inches. Find the total amount of rain the city had over the 4 months.

4. Find the cost of making 156 copies of a one-page form at $0.08 per copy.

5. In a state lottery prize, six people divided a prize of $30,528.18. How much was each person's share?

Answers

1. $149.60

2. 2.9 seconds

3. 17.03 inches

4. $12.48

5. $5,088.03

Changing Fractions to Decimals

A fraction can be converted to an equivalent decimal. For example $\frac{1}{4} = 0.25$. When a fraction is converted to a decimal, it will be in one of two forms: a **terminating decimal** or a **repeating decimal**.

To change a fraction to a decimal, divide the numerator by the denominator.

 EXAMPLE

Change $\frac{7}{8}$ to a decimal.

SOLUTION

$$
\begin{array}{r}
.875 \\
8\overline{)7.000} \\
\underline{6\ 4} \\
60 \\
\underline{56} \\
40 \\
\underline{40} \\
0
\end{array}
$$

Hence, $\frac{7}{8} = 0.875$.

EXAMPLE

Change $\dfrac{5}{16}$ to a decimal.

SOLUTION

$$
\begin{array}{r}
.3125 \\
16\overline{)5.0000} \\
\underline{4\ 8} \\
20 \\
\underline{16} \\
40 \\
\underline{32} \\
80 \\
\underline{80} \\
0
\end{array}
$$

Hence, $\dfrac{5}{16} = 0.3125$.

EXAMPLE

Change $\dfrac{7}{11}$ to a decimal.

SOLUTION

$$
\begin{array}{r}
.6363 \\
11\overline{)7.0000} \\
\underline{6\ 6} \\
40 \\
\underline{33} \\
70 \\
\underline{66} \\
40 \\
\underline{33} \\
6
\end{array}
$$

The quotient keeps repeating.

Hence, $\dfrac{7}{11} = 0.6363\ldots$

The repeating decimal can be written as $0.\overline{63}$.

EXAMPLE

Change $\dfrac{1}{6}$ to a decimal.

SOLUTION

$$
\begin{array}{r}
.166 \\
6\overline{)1.000} \\
\underline{6} \\
40 \\
\underline{36} \\
40 \\
\underline{36} \\
4
\end{array}
$$

Hence, $\dfrac{1}{6} = 0.166\ldots$ or $0.1\overline{6}$.

A mixed number can be changed to a decimal by first changing it to an improper fraction and then dividing the numerator by the denominator.

EXAMPLE

Change $8\dfrac{3}{5}$ to a decimal.

SOLUTION

$$
8\dfrac{3}{5} = \dfrac{43}{5} \qquad
\begin{array}{r}
8.6 \\
5\overline{)43.0} \\
\underline{40} \\
3\,0 \\
\underline{3\,0} \\
0
\end{array}
$$

Hence, $8\dfrac{3}{5} = 8.6$.

Practice

Change each fraction on mixed number to a decimal.

1. $\dfrac{3}{4} =$

2. $\dfrac{5}{6} =$

3. $\dfrac{13}{20}$

4. $\dfrac{7}{12}$

5. $9\dfrac{2}{3}$

Answers

1. 0.75

2. $0.8\overline{3}$

3. 0.65

4. $0.58\overline{3}$

5. $9.\overline{6}$

Changing Decimals to Fractions

To change a terminating decimal to a fraction, drop the decimal point and place the digits to the right of the decimal in the numerator of a fraction whose denominator corresponds to the place value of the last digit in the decimal. Reduce the answer if possible.

EXAMPLE

Change 0.4 to a fraction.

SOLUTION

$$0.4 = \frac{4}{10} = \frac{2}{5}$$

Hence, $0.4 = \dfrac{2}{5}$.

EXAMPLE

Change 0.64 to a fraction.

SOLUTION

$$0.64 = \frac{64}{100} = \frac{16}{25}$$

Hence, $0.64 = \dfrac{16}{25}$.

EXAMPLE

Change 0.0028 to a fraction.

SOLUTION

$$0.0028 = \frac{28}{10,000} = \frac{7}{2,500}$$

Hence, $0.0028 = \dfrac{7}{2,500}$.

Practice

Change each decimal to a reduced fraction:

1. 0.3
2. 0.08
3. 0.55
4. 0.375
5. 0.0032

Answers

1. $\dfrac{3}{10}$

2. $\dfrac{2}{25}$

3. $\dfrac{11}{20}$

4. $\dfrac{3}{8}$

5. $\dfrac{8}{2,500}$

Changing a repeating decimal to a fraction requires a more complex procedure, and this procedure is beyond the scope of this book. However, the following Table 5-1 can be used for some common repeating decimals.

The numbers that include the fractions and their equivalent terminating or repeating decimals are called **rational numbers**. Remember that all whole numbers and integers can be written as fractions, so these numbers are also rational numbers.

TABLE 5-1   Common Repeating Decimals
$\dfrac{1}{12} = 0.08\overline{3}$
$\dfrac{1}{6} = 0.1\overline{6}$
$\dfrac{1}{3} = 0.\overline{3}$
$\dfrac{5}{12} = 0.41\overline{6}$
$\dfrac{7}{12} = 0.58\overline{3}$
$\dfrac{2}{3} = 0.\overline{6}$
$\dfrac{5}{6} = 0.8\overline{3}$
$\dfrac{11}{12} = 0.91\overline{6}$

Fractions and Decimals

Sometimes a problem contains both fractions and decimals.

To solve a problem with a fraction and a decimal, either change the fraction to a decimal or change the decimal to a fraction and then perform the operation.

EXAMPLE

Add $\dfrac{1}{4} + 0.45$.

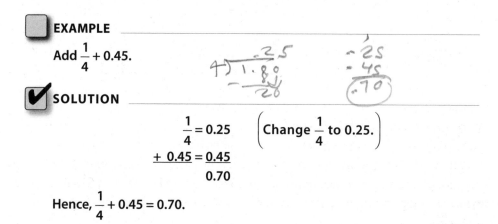

SOLUTION

$$\dfrac{1}{4} = 0.25 \qquad \left(\text{Change } \dfrac{1}{4} \text{ to 0.25.} \right)$$

$$+\ \underline{0.45 = 0.45}$$

$$0.70$$

Hence, $\dfrac{1}{4} + 0.45 = 0.70$.

 **ALTERNATE SOLUTION**

$$0.45 = \frac{45}{100} = \frac{9}{20} \qquad \left(\text{Change 0.45 to } \frac{9}{20}. \right)$$

$$\frac{1}{4} + \frac{9}{20} = \frac{5}{20} + \frac{9}{20}$$

$$= \frac{14}{20}$$

$$= \frac{7}{10}$$

Hence, $\frac{1}{4} + 0.45 = 0.7$ or $\frac{7}{10}$.

EXAMPLE

Multiply $\frac{5}{8} \times 0.25$.

SOLUTION

$$\frac{5}{8} = 0.625$$

$$0.625 \times 0.25 = 0.15625$$

ALTERNATE SOLUTION

$$0.25 = \frac{1}{4}$$

$$\frac{5}{8} \times \frac{1}{4} = \frac{5}{32}$$

Hence, $\frac{5}{8} \times 0.25 = 0.15625$ or $\frac{5}{32}$.

Practice

Perform the indicated operation:

1. $\frac{3}{8} + 0.7$

2. $\frac{13}{20} \times 0.64$

3. $\dfrac{9}{10} - 0.75$

4. $\dfrac{2}{5} \div 0.8$

5. $\dfrac{7}{20} + 0.72$

Answers

1. 1.075 or $1\dfrac{3}{40}$

2. 0.416 or $\dfrac{52}{125}$

3. 0.15 or $\dfrac{3}{20}$

4. 0.5 or $\dfrac{1}{2}$

5. 1.07 or $1\dfrac{7}{100}$

Operations with Positive and Negative Decimals

When performing operations with decimals that are positive and negative, perform the operations using the rules given in Chap. 2.

 EXAMPLE

Add –9.73 + (–0.57).

 **SOLUTION**

$$
\begin{array}{r}
-9.73 \\
\underline{-0.57} \\
-10.30
\end{array}
$$

Recall that when two negative numbers are added, the answer is negative.

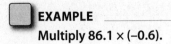

EXAMPLE
Multiply 86.1 × (−0.6).

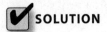

SOLUTION

$$\begin{array}{r} 86.1 \\ \times\ -0.6 \\ \hline -51.66 \end{array}$$

Since a negative number times a positive number is negative, the answer is −51.66.

Practice
Perform the indicated operations:

1. −1.6 − 0.8
2. 5.23 + (−9.375)
3. −6.3 × 8.1
4. −0.4896 ÷ (−0.8)
5. −12.62 − (−7.723)

Answers

1. −2.4
2. −4.145
3. −51.03
4. 0.612
5. −4.897

This chapter explained addition, subtraction, multiplication, and division of decimals. Conversions between fractions and decimals were also shown.

QUIZ

1. **In the number 5.731487, the place value of the 8 is:**
   A. hundredths
   B. thousandths
   C. ten-thousandths
   D. hundred-thousandths

2. **Name the number 0.047.**
   A. forty-seven hundredths
   B. forty-seven thousandths
   C. forty-seven ten-thousandths
   D. forty-seven hundred-thousandths

3. **Round 0.62827 to the nearest thousandth.**
   A. 0.63
   B. 0.628
   C. 0.629
   D. 0.6283

4. **Add 4.591 + 3.72 + 0.9812.**
   A. 9.2922
   B. 8.6815
   C. 9.3012
   D. 8.2357

5. **Subtract 71.2 − 37.615.**
   A. 32.617
   B. 33.823
   C. 33.585
   D. 32.168

6. **Multiply 0.061 × 3.2.**
   A. 3.261
   B. 3.139
   C. 0.2031
   D. 0.1952

7. **Multiply 0.006 × 0.004.**
   A. 0.000024
   B. 0.00024
   C. 0.0024
   D. 0.00000024

8. Divide 62.4672 ÷ 8.64.

   A. 72.3

   B. 0.0723

   C. 7.23

   D. 0.723

9. Divide 2.5143 ÷ 51.

   A. 0.493

   B. 4.93

   C. 0.0493

   D. 0.00493

10. Arrange in order of smallest to largest: 0.6, 0.006, 6.6, 0.606.

   A. 6.6, 0.006, 0.606, 0.6

   B. 0.6, 0.606, 0.006, 6.6

   C. 0.606, 0.6, 6.6, 0.606

   D. 0.006, 0.6, 0.606, 6.6

11. A person cut three pieces of wire from a piece 36 inches long. The lengths were 6.25 inches, 8.1 inches, and 7.4 inches. How much wire was left?

   A. 13.75 inches

   B. 14.25 inches

   C. 21.75 inches

   D. 16.25 inches

12. A phone card charges $1.09 for the first 10 minutes and $0.11 for each minute after that. Find the cost of a 27-minute call.

   A. $2.96

   B. $2.19

   C. $4.06

   D. $2.98

13. Change $\dfrac{13}{16}$ to a decimal.

   A. 0.825

   B. 0.6375

   C. 1.203

   D. 0.8125

14. Change $\dfrac{11}{12}$ to a decimal.

   A. $0.9\overline{16}$

   B. $1.\overline{09}$

   C. $0.91\overline{6}$

   D. $0.\overline{916}$

15. **Change 0.62 to a reduced fraction.**

   A. $\dfrac{3}{5}$

   B. $\dfrac{31}{50}$

   C. $\dfrac{62}{100}$

   D. $\dfrac{31}{52}$

16. **Change 0.245 to a reduced fraction.**

   A. $\dfrac{49}{200}$

   B. $\dfrac{1}{5}$

   C. $\dfrac{6}{25}$

   D. $\dfrac{49}{100}$

17. **Add $\dfrac{5}{16} + 0.75$**

   A. 0.3125
   B. 0.37523
   C. 1.0625
   D. 0.5625

18. **Multiply $\dfrac{9}{10} \times 0.06$.**

   A. 0.55
   B. 0.6

   C. $\dfrac{27}{500}$

   D. $\dfrac{3}{2}$

19. **Subtract −3.176 − (−0.43).**

   A. −0.7476
   B. 0.7476
   C. 2.746
   D. −2.746

20. **Divide −0.85 ÷ (0.5).**
    A. 0.17
    B. −0.80
    C. 0.80
    D. −1.7

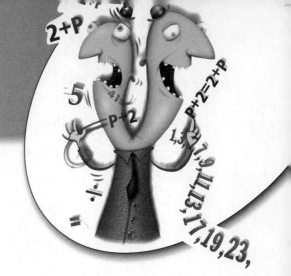

chapter **6**

Percent

This chapter explains the basic concepts of percent and the three types of percent problems. Percents are used extensively in everyday life. For example, when paying for a meal at a restaurant, you might include an 18% tip. Tax rates, in many cases, are expressed as percents. The state sales tax in Pennsylvania is 6%. In other areas, the results of surveys are expressed as percents. For example, 10–15% of the people born in the United States are left-handed.

CHAPTER OBJECTIVES

In this chapter, you will learn how to

- Change percents to decimals
- Change decimals to percents
- Change fractions to percents
- Change percents to fractions
- Solve the three types of percent problems
- Solve word problems using percents

Basic Concepts

Percent means hundredths. For example, 17% means $\frac{17}{100}$ or 0.17. Another way to think of 17% is to think of 17 equal parts out of 100 equal parts. See Fig. 6-1.

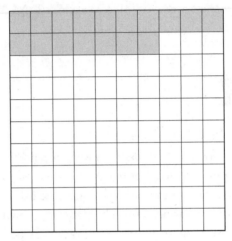

FIGURE 6-1

It should be pointed out that 100% means $\frac{100}{100}$ or 1.

Changing Percents to Decimals

To change a percent to a decimal, drop the percent sign and move the decimal point two places to the left.

▢ **EXAMPLE**
Write 5% as a decimal.

☑ **SOLUTION**

$$5\% = 0.05$$

Still Struggling

If there is no decimal point in the percent, it is at the end of the number. That is, 5% = 5.0%.

EXAMPLE

Write 72% as a decimal.

SOLUTION

$$72\% = 0.72$$

EXAMPLE

Write 271% as a decimal.

SOLUTION

$$271\% = 2.71$$

EXAMPLE

Write 22.8% as a decimal.

SOLUTION

$$22.8\% = 0.228$$

EXAMPLE

Write 0.6% as a decimal.

SOLUTION

$$0.6\% = 0.006$$

Practice
Write each percent as a decimal:

1. 43%  .43% = . 43
2. 7% .07% = .07
3. 153% 153% = 1.53
4. 14.3% 14.3% = .143
5. 0.2% 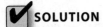 0.0 2% = .002

Answers

1. 0.43

2. 0.07

3. 1.53

4. 0.143

5. 0.002

Changing Decimals to Percents

To change a decimal to a percent, move the decimal point two places to the right and affix the percent sign.

MATH NOTE *If the decimal is located at the end of the number, do not write it.*

EXAMPLE
Write 0.84 as a percent.

SOLUTION

$$0.84 = 84\%$$

EXAMPLE
Write 0.09 as a percent.

SOLUTION

$$0.09 = 9\%$$

EXAMPLE

Write 1.45 as a percent.

SOLUTION

$$1.45 = 145\%$$

EXAMPLE

Write 0.556 as a percent.

SOLUTION

$$0.556 = 55.6\%$$

Still Struggling

If the decimal has only one place, make sure you add a zero after it when you change it to a percent. For example, 0.8 = 0.80 = 80%. The answer is not 8% since 8% = 0.08.

EXAMPLE

Write 0.0047 as a percent.

SOLUTION

$$0.0047 = 0.47\%$$

EXAMPLE

Write 9 as a percent.

SOLUTION

$$9 = 9.00 = 900\%$$

Still Struggling

Whenever you change percents to decimals or decimals to percents, you can remember which way to move the decimal point by using the first letter of each word, D and P. Since D comes before P alphabetically, move the point to the right when changing from decimal to percent and left when changing a percent to a decimal. It's the same direction as moving in the alphabet. To move from D to P, you move right. To move from P to D, you move left.

Practice
Write each decimal as a percent:

1. 0.02 $0.02. = 2\%$
2. 0.77 $0.77 = 77\%$
3. 0.225 $0.225 = 22.5\%$
4. 0.0011 $0.0011 = 0.11\%$
5. 5 $5.00\% = 500\%$

Answers

1. 2%
2. 77%
3. 22.5%
4. 0.11%
5. 500%

Changing Fractions to Percents

To change a fraction to a percent, change the fraction to a decimal (i.e., divide the numerator by the denominator) then move the decimal two places to the right and affix the percent sign.

 **EXAMPLE**

Write $\dfrac{3}{8}$ as a percent.

SOLUTION

$$
\begin{array}{r}
.375 = 37.5\% \\
8\overline{)3.000} \\
\underline{2\,4} \\
60 \\
\underline{56} \\
40 \\
\underline{40} \\
0
\end{array}
$$

 **EXAMPLE**

Write $\dfrac{4}{5}$ as a percent.

SOLUTION

$$
\begin{array}{r}
.8 = 80\% \\
5\overline{)4.0} \\
\underline{4\,0} \\
0
\end{array}
$$

EXAMPLE

Write $\dfrac{1}{4}$ as a percent.

SOLUTION

$$
\begin{array}{r}
.25 = 25\% \\
4\overline{)1.00} \\
\underline{8} \\
20 \\
\underline{20} \\
0
\end{array}
$$

EXAMPLE

Write $1\frac{2}{5}$ as a percent.

SOLUTION

$$1\frac{2}{5} = \frac{7}{5}$$

$$\begin{array}{r} 1.4 = 140\% \\ 5\overline{)7.0} \\ \underline{5} \\ 2\,0 \\ \underline{2\,0} \\ 0 \end{array}$$

EXAMPLE

Write $\frac{7}{12}$ as a percent.

SOLUTION

$$\begin{array}{r} .583 = 58.\overline{3}\% \\ 12\overline{)7.000} \\ \underline{6\,0} \\ 1\,00 \\ \underline{96} \\ 40 \\ \underline{36} \\ 4 \end{array}$$

Practice

Write each fraction as a percent:

1. $\frac{5}{8}$

2. $\frac{9}{16}$

3. $\frac{7}{50}$

4. $8\frac{3}{4}$

5. $\frac{11}{12}$

Answers

1. 62.5%

2. 56.25%

3. 14%

4. 875%

5. $91.\overline{6}\%$

Changing Percents to Fractions

To change a percent to a fraction, write the numeral in front of the percent sign as the numerator of a fraction whose denominator is 100. Reduce or simplify the fraction if possible.

EXAMPLE

Write 35% as a fraction.

SOLUTION

$$35\% = \frac{35}{100} = \frac{7}{20}$$

EXAMPLE

Write 20% as a fraction.

SOLUTION

$$20\% = \frac{20}{100} = \frac{1}{5}$$

EXAMPLE

Write 8% as a fraction.

SOLUTION

$$8\% = \frac{8}{100} = \frac{2}{25}$$

EXAMPLE

Write 155% as a fraction.

 **SOLUTION**

$$155\% = \frac{155}{100} = 1\frac{11}{20}$$

Practice

Write each percent as a fraction:

1. 6% $\frac{6 \div 2}{100 \div 2} = \frac{3}{50}$

2. 15% $\frac{15 \div 5}{100 \div 5} = \frac{3}{20}$

3. 185% $\frac{185 \div 5}{100 \div 5} = \frac{37}{20}$ $1\frac{17}{20}$

4. 70% $\frac{70 \div 10}{100 \div 10} = \frac{7}{10}$

5. 93% $\frac{93 \div 2}{100}$

Answers

1. $\frac{3}{50}$

2. $\frac{3}{20}$

3. $1\frac{17}{20}$

4. $\frac{7}{10}$

5. $\frac{93}{100}$

Three Types of Percent Problems

There are three basic types of percent problems and there are several different methods that can be used to solve these problems. The circle method will be used in this chapter. The equation method will be used in the next chapter, and finally, the proportion method will be used in Chap. 8. A percent problem has three values: the base (B) or whole, the rate (R) or percent, and the part (P). For example, if a class consisted of 20 students, 5 of whom were absent today, the base or whole would be 20, the part would be 5, and the rate or percent of students who were absent would be $\dfrac{5}{20} = 25\%$.

Draw a circle and place a horizontal line through the center and a vertical line halfway down in the center also. In the top section, write the word "is." In the lower left section write the percent sign, and in the lower right section write the word "of." In the top section place the part (P). In the lower left section place the rate (R) or percent number, and in the lower right section place the base (B). One of these three quantities will be unknown. See Fig. 6-2.

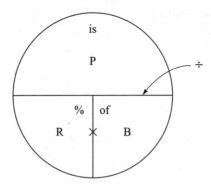

FIGURE 6-2

If you are given the two bottom numbers, multiply them to get the top number. That is, $P = R \times B$. If you are given the top number and one of the bottom numbers, divide to find the other number. That is $R = \dfrac{P}{B}$ or $B = \dfrac{P}{R}$. See Fig. 6-3.

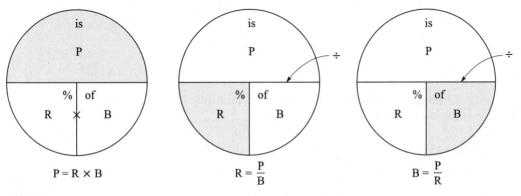

$$P = R \times B \qquad\qquad R = \dfrac{P}{B} \qquad\qquad B = \dfrac{P}{R}$$

FIGURE 6-3

Type 1: Finding the Part

Type 1 problems can be stated as follows:

- "Find 20% of 60."
- "What is 20% of 60?"
- "20% of 60 is what number?"

In Type 1 problems, you are given the base and the rate and are asked to find the part. From the circle: $P = R \times B$. Here, then, you change the percent to a decimal or fraction and multiply.

 EXAMPLE
Find 20% of 60.

 SOLUTION
Draw the circle and put 20% in the percent section and 60 in the "of" section.
See Fig. 6-4.

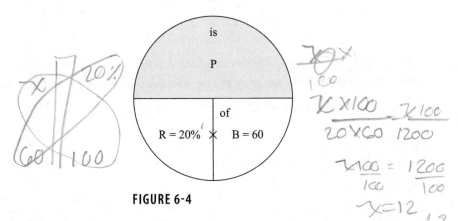

FIGURE 6-4

Change the percent to a decimal and multiply: $0.20 \times 60 = 12$.

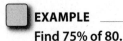 **EXAMPLE**
Find 75% of 80.

 SOLUTION
Draw the circle and put 75 in the percent section and 80 in the "of" section.
See Fig. 6-5.

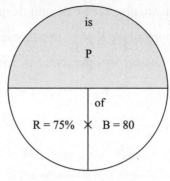

FIGURE 6-5

Change 75% to a decimal and multiply: 0.75 × 80 = 60.

MATH NOTE *Always change the percent to a decimal or fraction before multiplying or dividing.*

Practice

1. Find 48% of 133.
2. Find 60% of 250.
3. What is 125% of 48?
4. 37.5% of 64 is what number?
5. Find 15% of 360.

Answers

1. 63.84
2. 150
3. 60
4. 24
5. 54

Type 2: Finding the Rate

Type 2 problems can be stated as follows:

- "What percent of 16 is 10?"
- "10 is what percent of 16?"

In Type 2 problems, you are given the base and the part and are asked to find the rate or percent. The formula is $R = \dfrac{P}{B}$. In this case, divide the part by the base and then change the answer to a percent. See Fig. 6-6.

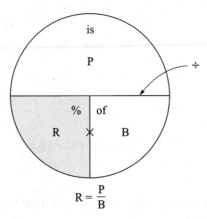

$$R = \dfrac{P}{B}$$

FIGURE 6-6

EXAMPLE

What percent of 25 is 15?

SOLUTION

Draw the circle and place 25 in the "of" section and 15 in the "is" section. See Fig. 6-7.

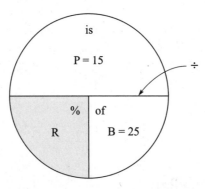

FIGURE 6-7

Then divide $\dfrac{15}{25} = 15 \div 25 = 0.60$. Change the decimal to a percent: $0.60 = 60\%$.

EXAMPLE

9 is what percent of 72?

SOLUTION

Draw the circle and put 9 in the "is" section and 72 in the "of" section. See Fig. 6-8.

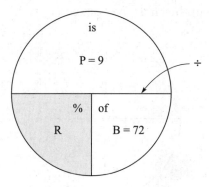

FIGURE 6-8

Then divide $\dfrac{9}{72} = 9 \div 72 = 0.125$. Change the decimal to a percent: $0.125 = 12.5\%$.

Practice

1. What percent of 10 is 8?

2. 35 is what percent of 120?

3. 36 is what percent of 40?

4. What percent of 60 is 12?

5. What percent of 90 is 80?

Answers

1. 80%

2. $29.1\overline{6}\%$

3. 90%

4. 20%

5. $88.\overline{8}\%$

Type 3: Finding the Base

Type 3 problems can be stated as follows:

- "16 is 20% of what number?"
- "20% of what number is 16?"

In Type 3 problems, you are given the rate and the part, and you are asked to find the base. From the circle, $B = \dfrac{P}{R}$. See Fig. 6-9.

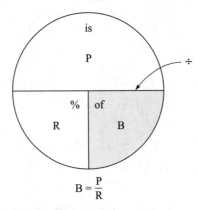

$$B = \frac{P}{R}$$

FIGURE 6-9

EXAMPLE

37% of what number is 222?

SOLUTION

Draw the circle and place 37 in the "percent" section and 222 in the "is" section. See Fig. 6-10.

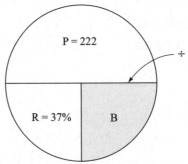

FIGURE 6-10

Change 37% to 0.37 and divide 222 ÷ 0.37 = 600.

EXAMPLE

165 is 55% of what number?

SOLUTION

Draw the circle and place 55 in the "percent" section and 165 in the "is" section. See Fig. 6-11.

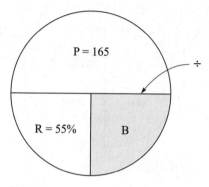

FIGURE 6-11

Change 55% to 0.55 and divide 165 ÷ 0.55 = 300.

Practice

1. 5% of what number is 60?

2. 350 is 70% of what number?

3. 54 is 40% of what number?

4. 80% of what number is 36.8?

5. 14.3% of what number is 9.295?

Answers

1. 1200

2. 500

3. 135

4. 46

5. 65

Word Problems

Percent word problems can be done using the circle. As stated previously, place the *rate* or *percent* in the lower left section. In the lower right section place the *base* or *whole*, and in the upper section place the *part*. See Fig. 6-12.

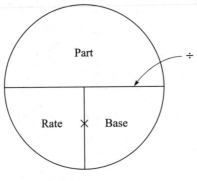

FIGURE 6-12

In order to solve a percent problem

- *Step 1: Read the problem.*
- *Step 2: Identify the base, rate (%), and part. One will be unknown.*
- *Step 3: Substitute the values in the circle.*
- *Step 4: Perform the correct operation. That is, either multiply or divide.*

EXAMPLE

On a test consisting of 60 problems, a student received a grade of 85%. How many problems did the student answer correctly?

SOLUTION

Place the rate, 85%, and the base, 60, into the circle. See Fig. 6-13.

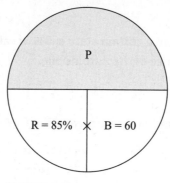

FIGURE 6-13

Change 85% to 0.85 and multiply $0.85 \times 60 = 51$. Hence, the student got 51 problems correct.

EXAMPLE

A baseball team won 36 of its 75 games. What percent of the games played did the team win?

SOLUTION

Place the 36 in the "part" section and the 75 in the "base" section of the circle. See Fig. 6-14.

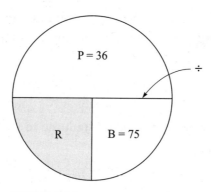

FIGURE 6-14

Then divide $\dfrac{36}{75} = 36 \div 75 = 0.48$. Hence, the team won 48% of its games.

EXAMPLE

The sales tax rate in a certain state is 5%. If sales tax on an automobile is $1,650, find the price of the automobile.

SOLUTION

Place the 5% in the "rate" section and the $1,650 in the "part" section. See Fig. 6-15.

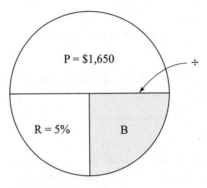

FIGURE 6-15

Change the 5% to 0.05 and divide. $1,650 ÷ 0.05% = $33,000. Hence, the price of the automobile is $33,000.

Another type of percent problem you will often see is the percent increase or percent decrease problem. In this situation, always remember that the *old* or *original* number is used as the base.

EXAMPLE

A suit that originally cost $200 was reduced to $120. What was the percent of the reduction?

SOLUTION

Find the amount of reduction: $200 − $120 = $80. Then place $80 in the "part" section and $200 in the "base" section of the circle. $200 is the base since it was the *original* price. See Fig. 6-16.

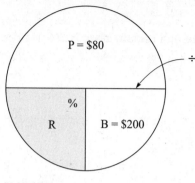

FIGURE 6-16

Divide 80 ÷ 200 = 0.40 = 40%. Hence, the cost was reduced by 40%.

Practice

1. A person bought an item at a 25% off sale. The discount was $32. What was the regular price?

2. A person saves $150 a month. If her annual income is $24,000 per year, what percent of her income is she saving?

3. A DVD player which originally sold for $80 was reduced $5. Find the percent of reduction.

4. In a psychology course which has 35 students, 20% of the students are mathematics majors. How many students are math majors?

5. A house sold for $72,500. If the salesperson receives a 3% commission, what was the amount of his commission?

6. A student correctly answered 45 of the 48 questions on a mathematics exam. What percent did she answer correctly?

7. A car dealer bought a car for $1,600 at an auction and then sold it for $4,000. What was the percent gain on the cost?

8. If the markup on a book is $6.75 and the book sold for $18, find the percent of the markup on the selling price.

9. The sales tax rate in a certain state is 6%. How much tax would be charged on a purchase of $53.98?

10. An automobile service station inspected 310 vehicles, and 80% of them passed. How many vehicles passed the inspection?

Answers

1. $128
2. 7.5%
3. 6.25%
4. 7
5. $2,175
6. 93.75%
7. 150%
8. 37.5%
9. $3.24
10. 248

In this chapter, the concept of percent was explained. There are three basic types of percent problems, and these can be solved by using the circle method. Two other methods of solving percent problems will be explained in later chapters.

QUIZ

1. **Write 9% as a decimal.**
   A. 0.009
   B. 0.09
   C. 0.9
   D. 9.0

2. **Write 22.2% as a decimal.**
   A. 0.0222
   B. 2.22
   C. 0.222
   D. 22.2

3. **Write 145% as a decimal.**
   A. 1.45
   B. 0.145
   C. 0.00145
   D. 145

4. **Write 0.27 as a percent.**
   A. 0.27%
   B. 0.0027%
   C. 2700%
   D. 27%

5. **Write 0.634 as a percent.**
   A. 63.4%
   B. 0.00634%
   C. 6.34%
   D. 634%

6. **Write 8 as a percent.**
   A. 0.8%
   B. 800%
   C. 80%
   D. 8%

7. **Write $\frac{3}{10}$ as a percent.**
   A. 3%
   B. 30%
   C. 300%
   D. 0.3%

8. Write $\frac{4}{9}$ as a percent.

   A. $4.\overline{4}\%$

   B. $0.\overline{4}\%$

   C. 444%

   D. $44.\overline{4}\%$

9. Write $6\frac{3}{4}$ as a percent.

   A. 6.75%

   B. 67.5%

   C. 675%

   D. 0.675%

10. Write 84% as a reduced fraction.

    A. $\frac{21}{25}$

    B. $1\frac{4}{25}$

    C. $\frac{1}{84}$

    D. $\frac{84}{100}$

11. Write 8% as a reduced fraction.

    A. $\frac{8}{50}$

    B. $\frac{5}{12}$

    C. $\frac{4}{5}$

    D. $\frac{2}{25}$

12. Write 325% as a reduced fraction or mixed number.

    A. $\frac{4}{13}$

    B. $3\frac{3}{4}$

    C. $3\frac{1}{4}$

    D. $\frac{4}{15}$

13. **Find 75% of 250.**
   A. 187.5
   B. 18.75
   C. 1875
   D. 18750

14. **8 is what percent of 25?**
   A. 32%
   B. 8%
   C. 80%
   D. 3.2%

15. **4% of what number is 60?**
   A. 15
   B. 240
   C. 1500
   D. 150

16. **38% of 114 is what number?**
   A. 300
   B. 43.32
   C. 30
   D. 4332

17. **A person bought a motorcycle for $8,000 and made a 25% down payment. How much was the down payment?**
   A. $320
   B. $200
   C. $3,200
   D. $2,000

18. **If the sales tax rate is 7% and the sales tax on an item is $35, find the cost of the item.**
   A. $500
   B. $24.50
   C. $0.05
   D. $5.00

19. **A person earned a commission of $208.80 on an item that sold for $2,320. Find the rate of the commission.**
   A. 5%
   B. 9%
   C. 8%
   D. 10%

20. A salesperson sold 4 chairs for $85 each and a table for $60. If the commission rate is 7.5%, find the person's commission.

    A. $4.50
    B. $22.50
    C. $30.00
    D. $10.88

chapter 7

Expressions and Equations

This chapter explains the basic concepts of algebraic expressions, formulas, and equations. These concepts can be used to solve a variety of problems. Formulas are used in all areas of mathematics and science. In Chap. 9, you will see how formulas can be used to find the perimeters, areas, and volumes of geometric figures. This chapter also shows how to solve percent problems using equations.

CHAPTER OBJECTIVES

In this chapter, you will learn how to

- Evaluate algebraic expressions
- Use the distributive property
- Combine like terms
- Remove parentheses and combine like terms
- Evaluate formulas
- Solve simple equations
- Solve more difficult equations
- Represent word statements mathematically
- Solve word problems using equations
- Solve percent problems using equations

Basic Concepts

In algebra, letters are used as variables. A **variable** can assume values of numbers. Numbers are called **constants**.

MATH NOTE *In some cases, a letter may represent a specific constant. As you will see in Chap. 9, the Greek letter pi (π) represents a constant.*

An **algebraic expression** consists of variables, constants, operation signs, and grouping symbols. In the algebraic expression 3x, the "3" is a constant and the "x" is a variable. When no sign is written between a number and a variable or between two or more variables, multiplication is implied. Hence the expression "3x" means "3 times x or to multiply 3 by the value of x. The expression abc means a times b times c or a × b × c. 3x means the same as (x)3, but the expression x3 is not usually written in algebra. Letters in algebraic expressions are usually placed in alphabetical order.

The number before the variable is called the **numerical coefficient**. In the algebraic expression "3x," the 3 is the numerical coefficient. When the numerical coefficient is 1, it is usually not written and vice versa. Hence, xy means 1xy. Likewise, 1xy is usually written as xy. Also, –xy means –1xy.

An algebraic expression consists of one or more **terms**. A **term** is a number or variable, or a product or a quotient of numbers and variables. Terms are connected by + or − signs. For example, the expression 3x + 2y − 6 has three terms. The expression 8p + 2q has two terms, and the expression $6x^2y$ consists of one term.

Evaluating Algebraic Expressions

*In order to **evaluate** an algebraic expression, substitute the values of the variables in the expression and simplify using the order of operations.*

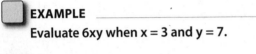 **EXAMPLE**

Evaluate 6xy when x = 3 and y = 7.

SOLUTION

$$6xy = 6(3)(7)$$
$$= 126$$

Hence, when x = 3 and y = 7, the value of the expression 6xy is 126.

**EXAMPLE**

Evaluate $-5x^3$ when $x = -4$.

SOLUTION

$$-5x^3 = -5(-4)^3$$
$$= -5(-64)$$
$$= 320$$

**EXAMPLE**

Evaluate $8(2x - y)$ when $x = 6$ and $y = -2$.

**SOLUTION**

$$8(2x - y) = 8[2(6) - (-2)]$$
$$= 8(12 + 2)$$
$$= 8(14)$$
$$= 112$$

Practice

Evaluate:

1. $32 - 3x$ when $x = -2$
2. $4x + 7y$ when $x = -6$ and $y = 3$
3. $(3x + 7)^2$ when $x = 4$
4. $5x - 4y + z$ when $x = 6$, $y = -5$, and $z = 3$
5. $6x^2 - 2y^2$ when $x = 4$ and $y = 10$

Answers

1. 38
2. –3
3. 361
4. 53
5. –104

The Distributive Property

An important property that is often used in algebra is called the **distributive property**.

For any numbers, a, b, and c, a(b + c) = ab + ac.

The distributive property states that when a sum of two numbers or expressions in parentheses is multiplied by a number outside the parentheses, the same result will occur if you multiply each number or expression inside the parentheses by the number outside the parentheses and add the results.

EXAMPLE

Multiply 3(4x + 7y).

12x + 24

SOLUTION

$$3(4x + 7y) = 3 \cdot 4x + 3 \cdot 7y$$
$$= 12x + 21y$$

EXAMPLE

Multiply 6(8x + y).

SOLUTION

$$6(8x + y) = 6 \cdot 8x + 6 \cdot y$$
$$= 48x + 6y$$

EXAMPLE

Multiply 5(7x − 6y).

SOLUTION

$$5(7x - 6y) = 5 \cdot 7x - 5 \cdot 6y$$
$$= 35x - 30y$$

The distributive property works for the sum or difference of three or more expressions in parentheses.

◻ **EXAMPLE**

Multiply 4(2a + 8b − 12c).

✔ **SOLUTION**

$$4(2a + 8b - 12c) = 4 \cdot 2a + 4 \cdot 8b - 4 \cdot 12c$$
$$= 8a + 32b - 48c$$

When a negative number appears outside the parentheses, make sure to change the signs of the terms inside the parentheses when multiplying.

◻ **EXAMPLE**

Multiply −5(3p − 9q − 6r).

✔ **SOLUTION**

$$-5(3p - 9q - 6r) = -5 \cdot 3p - (-5) \cdot 9q - (-5) \cdot 6r$$
$$= -15p + 45q + 30r$$

MATH NOTE   *The distributive property is also called the distributive property for multiplication over addition.*

Practice
Multiply:

1. 7(8x + 3y)
2. 6(d + 9)
3. 3(p + 2q − 8r)
4. −8(2x − 7y − 5z)
5. −2(6a − 7b + 9c)

Answers

1. 56x + 21y
2. 6d + 54
3. 3p + 6q − 24r
4. −16x + 56y + 40z
5. −12a + 14b − 18c

TABLE 7-1	Identification of Like Terms
Like Terms	Unlike Terms
2x, –7x	2x, –7y
$-8x^2, 3x^2$	$-8x^2, 3x$
13xy, –7xy	13xy, –7xz
$5x^2y, 3x^2y$	$5x^2y, 3xy^2$
x, 4x	x, 4

Combining Like Terms

Like terms have the same variables and the same exponents of the variables. For example, 3x and –5x are like terms because they have the same variables whereas 5x and –8y are **unlike terms** because they have different variables. Table 7-1 will help you to identify like terms.

Like terms can be added or subtracted. For example, 4x + 2x = 6x.

To add or subtract like terms, add or subtract the numerical coefficients and use the same variables.

EXAMPLE
Add 4x + 10x.

SOLUTION

4x + 10x = (4 + 10)x   (Note: This is the distribution property in reverse.)
= 14x

EXAMPLE
Subtract 12y – 8y.

SOLUTION

$$12y - 8y = (12 - 8)y$$
$$= 4y$$

Adding or subtracting like terms is often called "combining like terms."

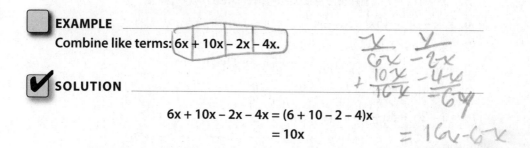

EXAMPLE
Combine like terms: 6x + 10x – 2x – 4x.

SOLUTION

$$6x + 10x - 2x - 4x = (6 + 10 - 2 - 4)x$$
$$= 10x$$

Unlike terms cannot be added or subtracted. When an expression has both like and unlike terms, it can be simplified by combining all like terms.

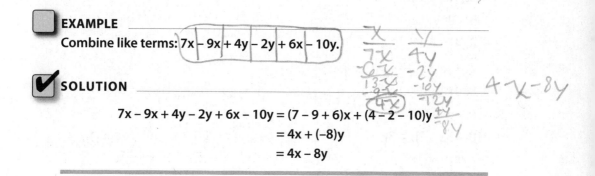

EXAMPLE
Combine like terms: 7x – 9x + 4y – 2y + 6x – 10y.

SOLUTION

$$7x - 9x + 4y - 2y + 6x - 10y = (7 - 9 + 6)x + (4 - 2 - 10)y$$
$$= 4x + (-8)y$$
$$= 4x - 8y$$

Still Struggling

An algebraic expression of the form a + (–b) is usually written as a – b.

Practice

For each expression, combine like terms:

1. 17x + 42x
2. 9y – 3y
3. x – 5x

4. $4x + 8x - 12x + 5x$

5. $-3y - 6y - 12y$

6. $32x - 9y + 15x - 3y + y$

7. $5x + 7y - 3 + 8x - 6y + 9$

8. $4y - x - y + 6x - x$

9. $9q + 6p - 7p + q$

10. $7a - 2b + 3c + 2b + 6a - 8b$

Answers

1. $59x$

2. $6y$

3. $-4x$

4. $5x$

5. $-21y$

6. $47x - 11y$

7. $13x + y + 6$

8. $4x + 3y$

9. $-p + 10q$

10. $13a - 8b + 3c$

Removing Parentheses and Combining Like Terms

Sometimes in algebra it is necessary to use the distributive property to remove parentheses and then combine terms. This procedure is also called "simplifying" an algebraic expression.

 EXAMPLE

Simplify $8(3x - 7) + 10$.

 SOLUTION

$8(3x - 7) + 10 = 24x - 56 + 10$ Remove parentheses

$\qquad\qquad\quad = 24x - 46$ Combine like terms

EXAMPLE

Simplify $-4(x - 8y) - 6x$.

 SOLUTION

$$-4(x - 8y) - 6x = -4x + 32y - 6x \quad \text{Remove parentheses}$$
$$= -10x + 32y \quad \text{Combine like terms}$$

Braces { } and brackets [ ] are also used as grouping symbols. Usually parentheses are placed inside brackets, and brackets are placed inside braces. For example, $5 + \{3 - [6 + 4(2 - 3)]\}$.

Always start with the innermost set of grouping symbols when simplifying expressions.

 EXAMPLE

Simplify $8 + \{21 - 7[4 + (8 \times 3)]\}$.

 **SOLUTION**

$$8 + \{21 - 7[4 + (8 \times 3)]\} = 8 + \{21 - 7[4 + 24]\}$$
$$= 8 + \{21 - 7[28]\}$$
$$= 8 + \{21 - 196\}$$
$$= 8 + \{-175\}$$
$$= -167$$

Practice

Simplify:

1. $9(x + 6) - 18$
2. $5(3x - 7y + 6) + 4x - 9$
3. $-2(6a + 7b) - 3a + 9b$
4. $4(a + 2b - 3c) + 6a - 7b$
5. $10\{6(x + 3y) - [3(2x - y)]\}$

Answers

1. $9x + 36$

2. $19x - 35y + 21$

3. $-15a - 5b$

4. $10a + b - 12c$

5. $210y$

Formulas

In mathematics and science, many problems can be solved using *formulas*. A **formula** is a mathematical statement of a relationship of two or more variables. For example, the relationship between Fahrenheit temperature and Celsius temperature can be expressed by the formula $F = \dfrac{9}{5}C + 32°$. Formulas can be evaluated in the same way as expressions are evaluated. That is, substitute the values of the variables in the formula and simplify using the order of operations.

 EXAMPLE

Find the Fahrenheit temperature corresponding to a Celsius temperature of $-15°$. Use the formula $F = \dfrac{9}{5}C + 32°$ where F = the Fahrenheit temperature and C = the Celsius temperature.

 SOLUTION

$$F = \frac{9}{5}C + 32°$$
$$= \frac{9}{5}(-15) + 32$$
$$= 9(-3) + 32$$
$$= -27 + 32$$
$$= 5$$

Hence, a Celsius temperature of $-15°$ is equivalent to a Fahrenheit temperature of $5°$.

EXAMPLE

Find the interest on a loan whose principal is $6,000 with a rate of 3% for 5 years. Use I = PRT or Interest = (Principal)(Rate)(Time).

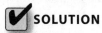**SOLUTION**

$$I = PRT$$
$$= (\$6{,}000)(0.03)(5)$$
$$= \$900$$

Practice

1. The amount of fence needed to enclose a rectangular garden is given by the formula $P = 2l + 2w$, where P is the perimeter, l is the length, and w is the width. Find the perimeter of a rectangular garden whose length is 18 feet and width is 12 feet.

2. The surface area of a cylindrical tank, excluding the top and bottom, is given by the formula $A = 2\pi rh$ where $\pi \approx 3.14$, r is the radius, and h is the height. Find the surface area of a cylindrical tank (excluding the top and bottom) when the radius is 15 inches and the height is 24 inches. The area is given in square inches.

3. Find the Celsius (C) temperature when the Fahrenheit (F) temperature is 68°. Use $C = \frac{5}{9}(F - 32°)$.

4. The resistance (R) in ohms in an electrical circuit is given by the formula $R = \frac{E}{I}$, where E is the electronic force in volts, and I is the current in amperes. In this case, E is 12 volts and I is 4 amperes. Find the resistance.

5. The amount of money (A) in a savings account is given by the formula $A = P + PRT$ where P is the principal, R is the rate, and T is the time in years. Find the amount of money you will have if you deposit $8,000 at 2% for 9 years.

Answers

1. 60 feet

2. 2,260.8 square inches

3. 20°C

4. 3 ohms

5. $9,440

Solving Simple Equations

An **equation** is a statement of equality of two algebraic expressions. For example, 3 + 2 = 5 is an equation. An equation can contain one or more variables. For example, x + 7 = 10 is an equation with one variable, x. If an equation has no variables, it is called a **closed** equation. Closed equations can be either true or false. For example, 6 + 8 = 14 is a closed true equation, whereas 3 + 5 = 6 is a false closed equation. Open equations, also called **conditional** equations, are neither true nor false. However, if a value for the variable is substituted in the equation and a closed true equation results, the value is called a **solution** or **root** of the equation. For example, when 3 is substituted for x in the equation x + 7 = 10, the resulting equation 3 + 7 = 10 is true, so 3 is called a solution or root of the equation. Finding the root of an equation is called **solving** the equation. The expression to the left of the equal sign in an equation is called the **left member** or **left side** of the equation. The expression to the right of the equal sign is called the **right member** or **right side** of the equation.

In order to solve an equation, it is necessary to transform the equation into a simpler equivalent equation with only the variable on one side and a constant or the root on the other side. There are four basic types of equations and four principles that are used to solve them. These principles do not change the nature of an equation. That means the simpler equivalent equation has the same solution as the original equation.

In order to **check** an equation, substitute the value of the solution or root for the variable in the original equation and see if a closed true equation results.

An equation such as x – 7 = 34 can be solved by using the **addition principle:** *the same number can be added to both sides of an equation without changing the nature of the equation.*

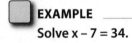**EXAMPLE**
Solve x – 7 = 34.

SOLUTION

$$x - 7 = 34$$
$$x - 7 + 7 = 34 + 7 \qquad \text{Add 7 to both sides}$$
$$x - 0 = 41$$
$$x = 41$$

Check:

$$x - 7 = 34$$
$$41 - 7 = 34$$
$$34 = 34$$

An equation such as $x + 8 = 14$ can be solved by the **subtraction principle:** *the same number can be subtracted from both sides of the equation without changing the nature of the equation.*

 EXAMPLE

Solve $x + 8 = 14$.

 **SOLUTION**

$$x + 8 = 14$$
$$x + 8 - 8 = 14 - 8 \qquad \textbf{Subtract 8 from both sides}$$
$$x + 0 = 6$$
$$x = 6$$

Check:

$$x + 8 = 14$$
$$6 + 8 = 14$$
$$14 = 14$$

An equation such as $3x = 27$ can be solved by using the **division principle:** *both sides of an equation can be divided by the same nonzero number without changing the nature of the equation.*

 EXAMPLE

Solve $3x = 27$.

 **SOLUTION**

$$3x = 27$$
$$\frac{\overset{1}{\cancel{3}}x}{\underset{1}{\cancel{3}}} = \frac{27}{3} \qquad \textbf{Divide both sides by 3}$$
$$x = 9$$

Check:

$$3x = 27$$
$$3(9) = 27$$
$$27 = 27$$

An equation such as $\dfrac{x}{6} = 13$ can be solved by using the **multiplication principle**: *both sides of an equation can be multiplied by the same nonzero number without changing the nature of the equation.*

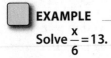 **EXAMPLE**

Solve $\dfrac{x}{6} = 13$.

 **SOLUTION**

$$\frac{x}{6} = 13$$

$$\frac{\overset{1}{\cancel{6}}}{1} \cdot \frac{x}{\underset{1}{\cancel{6}}} = 13 \cdot 6$$

$$x = 78$$

Check:

$$\frac{x}{6} = 13$$

$$\frac{78}{6} = 13$$

$$13 = 13$$

As you can see, there are four basic types of equations and four basic principles that are used to solve them. Before attempting to solve an equation, you should see what operation is being performed on the variable and then use the opposite principle to solve the equation. Addition and subtraction are opposite operations and multiplication and division are opposite operations.

 Still Struggling

It is important to understand that there is a difference between equations and expressions. Equations have equal signs and are to be solved for the variable. Expressions do not have an equal signs, and they are simplified.

Practice

Solve each of the following equations:

1. $x + 53 = 97$
2. $x - 17 = 41$
3. $6x = 42$
4. $14 + x = 3$
5. $x - 9 = -20$
6. $\dfrac{x}{12} = -4$
7. $98 = x + 62$
8. $-4x = 36$
9. $x + 4 = -12$
10. $\dfrac{x}{7} = -5$

Answers

1. 44
2. 58
3. 7
4. –11
5. –11
6. –48
7. 36
8. –9
9. –16
10. –35

Solving Equations Using Two Principles

Most equations require you to use more than one principle to solve them. These equations use the addition or subtraction principle first and then use the division principle.

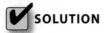

EXAMPLE

Solve $4x + 18 = 66$.

SOLUTION

$$4x + 18 = 66$$
$$4x + 18 - 18 = 66 - 18 \qquad \text{Subtract 18}$$
$$4x = 48$$
$$\frac{\overset{1}{\cancel{4}}x}{\underset{1}{\cancel{4}}} = \frac{48}{4} \qquad \text{Divide by 4}$$
$$x = 12$$

Check:

$$4x + 18 = 66$$
$$4(12) + 18 = 66$$
$$48 + 18 = 66$$
$$66 = 66$$

(handwritten:) $4x + 18 = 66$ $-18 \quad -18$ $4x = 48$ $\div \quad \div$ $x = 12$

EXAMPLE

Solve $2x + 14 = -6$.

SOLUTION

$$2x + 14 = -6$$
$$2x + 14 - 14 = -6 - 14 \qquad \text{Subtract 14}$$
$$2x = -20$$
$$\frac{\overset{1}{\cancel{2}}x}{\underset{1}{\cancel{2}}} = \frac{-20}{2} \qquad \text{Divide by 2}$$
$$x = -10$$

Check:

$$2x + 14 = -6$$
$$2(-10) + 14 = -6$$
$$-20 + 14 = -6$$
$$-6 = -6$$

EXAMPLE

Solve $15 - 5x = -70$.

SOLUTION

$$15 - 5x = -70$$
$$15 - 15 - 5x = -70 - 15 \quad \text{Subtract 15}$$
$$-5x = -85$$
$$\frac{-5x}{-5} = \frac{-85}{-5} \quad \text{Divide by } -5$$
$$x = 17$$

Check:

$$15 - 5x = -70$$
$$15 - 5(17) = -70$$
$$15 - 85 = -70$$
$$-70 = -70$$

Practice

Solve each equation:

1. $13x + 15 = 106$
2. $-2 + 7x = -51$
3. $18 + 6x = 48$
4. $9x - 3 = 33$
5. $62 - 5x = -33$

Answers

1. 7
2. -7
3. 5
4. 4
5. 19

Solving More Difficult Equations

When an equation has like terms on the same side, these terms can be combined first.

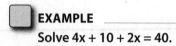 **EXAMPLE**

Solve $4x + 10 + 2x = 40$.

 **SOLUTION**

$$4x + 10 + 2x = 40$$
$$6x + 10 = 40 \qquad \text{Combine } 4x + 2x$$
$$6x + 10 - 10 = 40 - 10 \qquad \text{Subtract 10}$$
$$6x = 30$$

$$\frac{\overset{1}{\cancel{6}}x}{\underset{1}{\cancel{6}}} = \frac{30}{6} \qquad \text{Divide by 6}$$

$$x = 5$$

Check:

$$4x + 10 + 2x = 40$$
$$4(5) + 10 + 2(5) = 40$$
$$20 + 10 + 10 = 40$$
$$40 = 40$$

If an equation has parentheses, use the distributive property to remove parentheses.

 **EXAMPLE**

Solve $5(2x - 8) = 60$.

 **SOLUTION**

$$5(2x - 8) = 60$$
$$10x - 40 = 60 \qquad \text{Remove parentheses}$$
$$10x - 40 + 40 = 60 + 40 \qquad \text{Add 40}$$
$$10x = 100$$

$$\frac{\overset{1}{\cancel{10}}x}{\underset{1}{\cancel{10}}} = \frac{100}{10} \qquad \textbf{Divide by 10}$$

$$x = 10$$

Check:

$$5(2x - 8) = 60$$
$$5(2 \cdot 10 - 8) = 60$$
$$5(12) = 60$$
$$60 = 60$$

The procedure for solving equations in general is:

- *Step 1: Remove parentheses.*
- *Step 2: Combine like terms on each side of the equation.*
- *Step 3: Use the addition and/or subtraction principle to get the variables on one side and the constant terms on the other side.*
- *Step 4: Use the division principle to solve for x.*
- *Step 5: Check the equation.*

 EXAMPLE ⎯⎯⎯⎯⎯⎯⎯⎯⎯⎯⎯⎯⎯⎯⎯⎯⎯⎯⎯⎯⎯⎯⎯⎯⎯⎯⎯⎯⎯⎯
Solve $8(2x - 3) - 4x = 6x + 18$.

 **SOLUTION** ⎯⎯⎯⎯⎯⎯⎯⎯⎯⎯⎯⎯⎯⎯⎯⎯⎯⎯⎯⎯⎯⎯⎯⎯⎯⎯⎯⎯⎯

$8(2x - 3) - 4x = 6x + 18$	
$16x - 24 - 4x = 6x + 18$	**Remove parentheses**
$12x - 24 = 6x + 18$	**Combine like terms**
$12x - 6x - 24 = 6x - 6x + 18$	**Get variables on one side and the constants on the other side**
$6x - 24 = 18$	
$6x - 24 + 24 = 18 + 24$	
$6x = 42$	

$$\frac{\overset{1}{\cancel{6}}x}{\underset{1}{\cancel{6}}} = \frac{42}{6} \qquad \textbf{Divide by 6}$$

$$x = 7$$

Check:

$$8(2x - 3) - 4x = 6x + 18$$
$$8(2 \cdot 7 - 3) - 4 \cdot 7 = 6 \cdot 7 + 18$$
$$8(14 - 3) - 28 = 42 + 18$$
$$8(11) - 28 = 42 + 18$$
$$88 - 28 = 42 + 18$$
$$60 = 60$$

Practice

Solve each equation:

1. $5x + 9 = 2x + 54$
2. $3x - 8 = 7x + 12$
3. $x - 19 = 17 - 8x$
4. $12x - 8 = 10x + 20$
5. $11x + 49 = 4x$
6. $24 = 4(x - 2)$
7. $5(2x - 6) = 3x + 12$
8. $9x - 10 = -5 - 2(x + 8)$
9. $-6(x + 6) = 0$
10. $4(3x + 6) = 6(x + 4)$

Answers

1. 15
2. −5
3. 4
4. 14
5. −7
6. 8
7. 6
8. −1
9. −6
10. 0

Algebraic Representation of Statements

In order to solve word problems in algebra, it is necessary to translate the verbal statements into algebraic expressions. An unknown can be designated by a variable, usually x. For example, the statement "four times a number plus ten" can be written algebraically as "4x + 10."

EXAMPLE

Write each statement in symbols:

1. eight times the sum of a number and 3
2. the product of 12 and a number
3. six subtracted from three times a number
4. a number divided by six
5. the sum of a number and 14

SOLUTION

1. 8(x + 3)
2. 12x
3. 3x – 6
4. x ÷ 6
5. x + 14

In solving word problems it is also necessary to represent two quantities using the same variable. For example, if the sum of two numbers is 10, and one number is x, the other number would be 10 – x. The reason is that given one number, say 7, you can find the other number by subtracting 10 – 7 to get 3. Another example—suppose you are given two numbers and the condition that one number is twice as large as the other. How would you represent the two numbers? The smaller number would be x, and since the second number is twice as large, it can be represented as 2x.

EXAMPLE

Represent algebraically two numbers such that one number is five more than twice the other number.

SOLUTION

Let x = one number

Let 2x + 5 = the other number

EXAMPLE

Represent algebraically two numbers such that the difference between the two numbers is 9.

SOLUTION

Let x = one number

Let x – 9 = the other number

Practice

1. Represent algebraically two numbers such that one number is one-third of another number.

2. Represent algebraically two numbers so that one number is eight more than another number.

3. Represent algebraically two numbers so that their sum is 31.

4. Represent algebraically two numbers such that one number is seven more than three times the other number.

5. Represent algebraically two numbers such that one number is six times larger than another number.

Answers

1. x, $\frac{1}{3}$x, or x, 3x
2. x, x + 8
3. x, 31 – x
4. x, 3x + 7
5. x, 6x

Word Problems

Equations can be used to solve word problems in algebra.

After reading the problem:

- *Step 1: Represent one of the unknown quantities as x and the other unknown using algebraic expressions in terms of x.*
- *Step 2: Write an equation using the unknown quantities.*
- *Step 3: Solve the equation for x and then find the other unknowns.*
- *Step 4: Check the answers.*

EXAMPLE

One number is 12 more than another number. The sum of the two numbers is 88. Find the numbers.

 SOLUTION

Step 1: Let x = the smaller number

Let x + 12 = the larger number

Step 2: x + x + 12 = 88

Step 3: 2x + 12 − 12 = 88 − 12

$$2x = 76$$

$$\frac{\overset{1}{\cancel{2}}x}{\underset{1}{\cancel{2}}} = \frac{76}{2}$$

$$x = 38$$

$$x + 12 = 50$$

Hence, one number is 38 and the other number is 50.

Step 4: Check: 38 + 50 = 88

 EXAMPLE

A person cuts a piece of ribbon 84 inches long into two pieces. One piece was three times as long as the other. Find the lengths of the pieces.

 SOLUTION

Step 1: Let x = the length of the smaller piece

Let 3x = the length of the larger piece

Step 2: x + 3x = 84

Step 3: x + 3x = 84

$$4x = 84$$

$$\frac{\overset{1}{\cancel{4}}x}{\underset{1}{\cancel{4}}} = \frac{84}{4}$$

$$x = 21 \text{ inches}$$

$$3x = 3 \cdot 21 = 63 \text{ inches}$$

Hence, one piece is 21 inches long and the other piece is 63 inches long.

Step 4: Check: 21 inches + 63 inches = 84 inches

EXAMPLE

$800 is to be divided among three people. The first person gets three times as much as the second; the third person gets $200. Find the amount each person received.

SOLUTION

Step 1: Let x = the amount the second person receives

Let 3x = the amount the first person receives

Step 2: x + 3x + 200 = 800

$$4x + 200 = 800$$

$$4x = 600$$

$$\frac{\overset{1}{\cancel{4}}x}{\underset{1}{\cancel{4}}} = \frac{600}{4}$$

$$x = \$150$$

$$3x = 3(150) = \$450$$

Hence, the first person received $450, the second person received $150, and the third person received $200.

Step 4: Check: $150 + $450 + $200 = $800

MATH NOTE *Remember to change a percent into a fraction or decimal before multiplying or dividing.*

Practice

1. A computer is worth four times what the printer that comes with it is worth. If the total price of the computer and the printer is $680, find the value of each item.

2. The difference between two numbers is 11. If the sum of the two numbers is 39, find the numbers.

3. A piece of ribbon 30 inches long is cut into two pieces. One piece is 6 inches longer than twice the other piece. Find the lengths of both the pieces.

4. A father is four times as old as his son. If the sum of their ages is 55, find each one's age.

5. The cost of a book, including a 6% tax is $51.94. Find the cost of the book if the tax is not included.

8. Find the wing loading (L) in pounds per square foot of an airplane when the gross weight (W) is 6,300 pounds and the wing area (A) is 450 square feet. Use $L = \dfrac{W}{A}$.

   A. 14 pounds per square foot

   B. 16 pounds per square foot

   C. 5 pounds per square foot

   D. 12 pounds per square foot

9. Find the distance (d) in feet of an object that falls when g = 32 and t = 5 seconds. Use $d = \dfrac{1}{2}gt^2$.

   A. 96 feet

   B. 160 feet

   C. 80 feet

   D. 400 feet

10. Solve x + 51 = 43.

   A. 94

   B. −8

   C. −94

   D. 8

11. Solve 9x = −108.

   A. 15

   B. −972

   C. −12

   D. −99

12. Solve x − 43 = 106.

   A. 63

   B. −149

   C. 149

   D. −63

13. Solve $\dfrac{x}{5} = 60$.

   A. 300

   B. 12

   C. 65

   D. 55

14. Solve 7x + 9 = −47.

   A. 8

   B. −3

   C. 5

   D. −8

15. Solve $3(6x - 10) = 60$.
   A. 18
   B. 5
   C. 10
   D. −18

16. Solve $7x - 3 - 2x = 24 - 4x$.
   A. 5
   B. 6
   C. 3
   D. −6

17. Represent algebraically two numbers so that one is four more than six times the other.
   A. $x, 6(x + 4)$
   B. $x, 4x$
   C. $x, 4x + 6$
   D. $x, 6x + 4$

18. The sum of two numbers is 45. If one number is five less than the other number, find the numbers.
   A. 20, 25
   B. 10, 35
   C. 8, 37
   D. 16, 21

19. What percent of 48 is 16? Solve using an equation.
   A. 300%
   B. 30%
   C. 33.3%
   D. 133%

20. 45% of what number is 405? Solve using an equation.
   A. 182.25
   B. 900
   C. 182.75
   D. 9

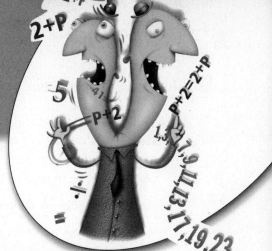

chapter 8

Ratio and Proportion

This chapter explains ratios and proportions. A ratio can be used to compare two items. For example, in Washington, D.C., the ratio of households without an automobile to those who own an automobile is 1 to 9. This means that, on average, for every 10 households in the metropolitan area of Washington, D.C., one household does not own an automobile, and nine households own an automobile. A proportion consists of two equal ratios. Many real-life problems can be solved using proportions. These types of problems are also explained in this chapter.

CHAPTER OBJECTIVES

In this chapter, you will learn how to

- Write ratios as fractions
- Solve proportions
- Solve word problems using proportions
- Solve percent problems using proportions

Ratio

A **ratio** is a comparison of two numbers. For example, consider the following statement, "About 14 out of every 100 people in the United States do not have health insurance."

A ratio can be expressed by a fraction or by using a colon. In the preceding example, the ratio $\dfrac{14}{100}$ is the same as 14 to 100 or 14:100. Fractions are usually reduced to lowest terms, so the ratio becomes $\dfrac{7}{50}$ or 7:50.

It is important to understand that whatever number comes first in the ratio statement is placed in the numerator of the fraction and whatever number comes second in the ratio statement is placed in the denominator of the fraction. In general, the ratio of a to b is written as $\dfrac{a}{b}$.

 EXAMPLE

Find the ratio of 9 to 12.

 SOLUTION

$$9 \text{ to } 12 = \frac{9}{12} = \frac{3}{4}$$

Generally, when the ratio of two numbers is given, the numbers must be in the same units; however, when the units are different, the units must be expressed. For example, if you drive 60 miles per hour, the ratio is given as $\dfrac{60 \text{ miles}}{1 \text{ hour}}$.

 EXAMPLE

If 12 oranges cost $3.00, find the ratio of oranges to cost.

 SOLUTION

$$\frac{12 \text{ oranges}}{3 \text{ dollars}} = \frac{4 \text{ oranges}}{1 \text{ dollars}} = \frac{4 \text{ oranges}}{\$1.00}$$

Make sure that when the units can be converted, the ratio uses the same units in the numerator and denominator. For example, the ratio of 18 eggs to 3 dozen eggs is not 18:3 since 3 dozen eggs is 3 × 12 or 36 eggs, so the true ratio is 18:36 or 1:2.

Practice

1. Find the ratio of 20 to 30.

2. Find the ratio of 15 to 6.

3. Find the ratio of 60 miles to 3 hours.

4. Find the ratio of 4 yards to 80 cents.

5. Find the ratio of one dime to one quarter.

Answers

1. $\dfrac{2}{3}$

2. $\dfrac{5}{2}$

3. $\dfrac{20 \text{ miles}}{1 \text{ hour}}$

4. $\dfrac{1 \text{ yard}}{20 \text{ cents}}$

5. $\dfrac{2}{5}$

Proportion

A **proportion** is a statement of equality of two ratios. For example, $\dfrac{3}{4} = \dfrac{6}{8}$ is a proportion. Proportions can also be expressed using a colon, as $3:4 = 6:8$.

A proportion consists of four terms. Usually, it is necessary to find the value of one term given the other three terms. For example, $\dfrac{5}{8} = \dfrac{x}{24}$. In order to do this, it is necessary to use a basic rule of proportions:

$$\frac{a}{b} = \frac{c}{d} \quad \text{if } a \cdot d = b \cdot c$$

In other words, $\dfrac{3}{4} = \dfrac{6}{8}$ if $3 \times 8 = 4 \times 6$ or $24 = 24$. This is called **cross multiplication.**

In order to find the unknown in a proportion, cross multiply and then solve the equation for the unknown. That is, divide by the number that is in front of the variable.

 **EXAMPLE**

Find the value of x:

$$\frac{x}{6} = \frac{24}{36}$$

SOLUTION

$$\frac{x}{6} = \frac{24}{36}$$

$$36x = 6 \cdot 24 \qquad \text{Cross multiply}$$

$$\frac{\cancel{36}x}{\cancel{36}} = \frac{144}{36} \qquad \text{Divide}$$

$$x = 4$$

 **EXAMPLE**

Find the value of x:

$$\frac{14}{42} = \frac{x}{3}$$

SOLUTION

$$\frac{14}{42} = \frac{x}{3}$$

$$14 \cdot 3 = 42x \qquad \text{Cross multiply}$$

$$42 = 42x$$

$$\frac{42}{42} = \frac{\cancel{42}x}{\cancel{42}} \qquad \text{Divide}$$

$$1 = x$$

EXAMPLE

Find the value of x:

$$\frac{15}{x} = \frac{75}{32}$$

✔ SOLUTION

$$\frac{15}{x} = \frac{75}{32}$$

$15 \cdot 32 = 75x$ **Cross multiply**

$480 = 75x$

$$\frac{480}{75} = \frac{75x}{75}$$ **Divide by 75**

$x = 6.4$

Practice

Find the value of x:

1. $\dfrac{5}{8} = \dfrac{20}{x}$

2. $\dfrac{8}{x} = \dfrac{32}{150}$

3. $\dfrac{x}{4} = \dfrac{25}{200}$

4. $\dfrac{81}{180} = \dfrac{30}{x}$

5. $\dfrac{x}{7} = \dfrac{6}{25}$

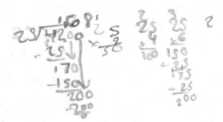

Answers

1. 32

2. 37.5

3. 0.5

4. $66.\overline{6}$

5. 1.68

Word Problems

In order to solve word problems using proportions:

- *Step 1: Read the problem.*
- *Step 2: Identify the ratio statement.*

- Step 3: Set up the proportion.
- Step 4: Solve for x.

In problems involving proportions, there will always be a ratio statement. It is important to find the ratio statement and then set up the proportion using the ratio statement. Be sure to keep the identical units in the numerators and denominators of the fractions in the proportion.

 EXAMPLE

A recipe for 4 servings calls for 6 tablespoons of shortening. If the cook wants to make 10 servings, how many tablespoons of shortening will be needed?

 SOLUTION

The ratio statement is:

$$\frac{6\,\text{tablespoons}}{4\,\text{servings}}$$

so the proportion is:

$$\frac{6\,\text{tablespoons}}{4\,\text{servings}} = \frac{x\,\text{tablespoons}}{10\,\text{servings}}$$

Solving $\dfrac{6}{4} = \dfrac{x}{10}$

$$6 \cdot 10 = 4x$$

$$60 = 4x$$

$$\frac{60}{4} = \frac{\cancel{4}x}{\cancel{4}}$$

$$15 = x$$

Hence, 15 tablespoons of shortening will be needed to make 10 servings.

MATH NOTE   *Notice in the previous example that the numerators of the proportions have the same units, tablespoons, and the denominators have the same units, servings.*

EXAMPLE

If it takes 50 pounds of fertilizer to cover a lawn with an area of 2,500 square feet, how many pounds of fertilizer will it take to cover a lawn that is 3,800 square feet?

 SOLUTION

The ratio statement is:

$$\frac{50\,\text{pounds}}{2,500\,\text{square feet}}$$

The proportion is:

$$\frac{50\,\text{pounds}}{2,500\,\text{square feet}} = \frac{x\,\text{pounds}}{3,800\,\text{square feet}}$$

$$\frac{50}{2,500} = \frac{x}{3,800}$$

$$50 \cdot 3,800 = 2,500x$$

$$190,000 = 2,500x$$

$$\frac{190,000}{2,500} = \frac{2,500x}{2,500}$$

$$76 = x$$

Hence, it will take 76 pounds to cover 3,800 square feet.

Practice

1. A motorist traveled 375 miles and used 15 gallons of gasoline. How many gallons of gasoline will be needed to travel 500 miles?

2. If a person pays $8,200 taxes on a house worth $156,000, how much tax (at the same rate) would there be on a house worth $96,000?

3. Mario can save $800 in 4 months. If he plans a trip that will cost $3,250, how many months will it take him to save enough money to pay for the trip?

4. On a map, 4 inches represents 325 miles. What is the distance in miles between two cities that are $2\frac{1}{2}$ inches apart?

5. If 4 gallons of waterproofing will cover a deck that is 480 square feet, how many gallons will it take to cover a deck that is 600 square feet?

Answers

1. 20 gallons

2. $5,046.15

3. 16.25 months

4. 203.125 miles

5. 5 gallons

Solving Percent Problems Using Proportions

Recall that percent means part of 100, so a percent can be written as a ratio using 100 as the denominator of the fraction. For example, 43% can be written as $\dfrac{43}{100}$. Also recall that a percent problem has a base, B, a part, P, and a rate (percent) R, so a proportion can be set up for each type of percent problem as $\dfrac{R}{100} = \dfrac{P}{B}$.

 EXAMPLE
Find 64% of 52.

 SOLUTION

$$\frac{64}{100} = \frac{P}{52}$$

$$64 \cdot 52 = 100P$$

$$3328 = 100P$$

$$\frac{3328}{100} = \frac{\overset{1}{\cancel{100}}\,P}{\underset{1}{\cancel{100}}}$$

$$33.28 = P$$

 EXAMPLE
What percent of 75 is 60?

SOLUTION

$$\frac{R}{100} = \frac{60}{75}$$

$$75R = 60 \cdot 100$$

$$75R = 6000$$

$$\frac{\overset{1}{\cancel{75}}R}{\underset{1}{\cancel{75}}} = \frac{6000}{75}$$

$$R = 80, \text{ so the rate is } 80\%.$$

EXAMPLE

34% of what number is 136?

SOLUTION

$$\frac{34}{100} = \frac{136}{B}$$

$$34B = 136 \cdot 100$$

$$34B = 13,600$$

$$\frac{\overset{1}{\cancel{34}}B}{\underset{1}{\cancel{34}}} = \frac{13,600}{34}$$

$$B = 400$$

Practice

Solve each percent problem by using a proportion:

1. Find 24% of 120.

2. 9 is what percent of 25?

3. 25% of what number is 128?

4. 16 is what percent of 125?

5. Find 98% of 600.

6. 63% of what number is 268.38?

7. 20 is what percent of 32?

8. Find 18% of 90.

9. 20% of what number is 64?

10. 7 is what percent of 35?

Answers

1. 28.8
2. 36%
3. 512
4. 12.8%
5. 588
6. 426
7. 62.5%
8. 16.2
9. 320
10. 20%

In this chapter, the concepts of ratio and proportion were explained along with how to solve real-life problems using proportions. Along with the other methods of solutions, percent problems can be solved with proportions.

QUIZ

1. **Find the ratio of 3 to 8.**

    A. $\dfrac{8}{3}$

    B. $\dfrac{3}{11}$

    C. $\dfrac{3}{8}$ Ⓒ $\dfrac{3}{8}$

    D. $\dfrac{8}{11}$

2. **Find the ratio of 4 to 16.**

    A. 16:20
    B. 1:4 $\dfrac{4 \div 4}{16 \div 4} = \dfrac{1}{4}$ Ⓑ
    C. 4:1
    D. 4:20

3. **Find the ratio of 15 to 5.**

    A. $\dfrac{3}{1}$ $\dfrac{15 \div 5}{5 \div 5} = \dfrac{3}{1}$

    B. $\dfrac{1}{4}$

    C. $\dfrac{5}{15}$ Ⓐ

    D. $\dfrac{1}{3}$

4. **Find the ratio of 8 to 13.**

    A. 13:8 $\dfrac{8}{13}$
    B. 8:21
    C. 13:21   Ⓓ $\dfrac{8}{13}$
    D. 8:13

5. **Find the ratio of 150 to 90.**

    A. $\dfrac{3}{8}$ $\dfrac{150 \div 3}{90 \div 3}$ $\dfrac{50 \div 10}{30 \div 10} = \dfrac{5}{3}$

    B. $\dfrac{5}{3}$

    C. $\dfrac{5}{8}$ Ⓑ

    D. $\dfrac{3}{5}$

6. Find the value of x: $\dfrac{5}{9} = \dfrac{x}{27}$.

   A. 3
   B. 15
   C. 8
   D. 5

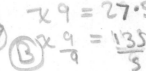

7. Find the value of x: $\dfrac{1}{8} = \dfrac{15}{x}$.

   A. 3
   B. 25
   C. 90
   D. 120

8. Find the value of x: $\dfrac{x}{8} = \dfrac{9}{12}$.

   A. 6
   B. 9
   C. 3
   D. 4

9. Find the value of x: $\dfrac{x}{5} = \dfrac{7}{15}$.

   A. 2.5
   B. 1.8
   C. $2.\overline{3}$
   D. $6.\overline{2}$

10. Find the value of x: $\dfrac{0.9}{0.15} = \dfrac{x}{0.5}$.

    A. 1.8
    B. 1.4
    C. 3.0
    D. 3.6

11. Find the value of x: $\dfrac{x}{0.6} = \dfrac{0.9}{0.2}$.

    A. 2.7
    B. 3.8
    C. 4.2
    D. 5.1

12. Find the value of x: $\dfrac{3}{x} = \dfrac{7}{9}$.

    A. $3\dfrac{1}{7}$

    B. $3\dfrac{6}{7}$

C. $7\dfrac{1}{6}$

D. $3\dfrac{1}{7}$

13. On a scale drawing, 5 inches represents 18 feet. How tall is a tree that is 3 inches high?

   A. 8 feet
   B. 9.6 feet
   C. 10.8 feet
   D. 12 feet

14. If 9 pounds of candy costs $7.20, how much will 14 pounds cost?

   A. $112.0
   B. $1.20
   C. $2.57
   D. $11.20

15. Betty saved $52 in 8 weeks. How long will it take her to save $312?

   A. 24 weeks
   B. 32 weeks
   C. 40 weeks
   D. 48 weeks

16. If a picture 4 inches wide and 6 inches long is to be enlarged so that the length is 30 inches, what would be the width of the enlarged picture?

   A. 20 inches
   B. 18 inches
   C. 15 inches
   D. 12 inches

17. If 25 feet of a certain type of cable weighs $2\dfrac{1}{4}$ pounds, how much will 40 feet of cable weigh?

   A. 2.5 pounds
   B. 3.6 pounds
   C. 4 pounds
   D. 4.75 pounds

18. What percent of 16 is 7? Solve using a proportion.

   A. 43.75%
   B. $16.\overline{6}$%
   C. 45%
   D. 52%

19. Solve using a proportion: 18% of 60 is what number?
    A. 8
    B. 10.8
    C. 12.4
    D. 15

20. Solve using a proportion: 18% of what number is 90?
    A. 5
    B. 50
    C. 500
    D. 5,000

chapter 9

Informal Geometry

This chapter explains the concepts of informal geometry. The various types of geometric figures are introduced. Then the formulas for finding the perimeters, areas, and volumes of the geometric figures are given. Many problems involving right triangles can be solved using the Pythagorean Theorem. The solutions to these types of problems are presented in this chapter.

CHAPTER OBJECTIVES

In this chapter, you will learn how to

- Identify the basic geometric figures
- Find perimeters of geometric figures
- Find areas of geometric figures
- Find volumes of geometric figures
- Solve word problems involving geometric figures
- Find the square root of a number
- Use the Pythagorean Theorem

Geometric Figures

The word "geometry" is derived from two Greek words meaning "earth measure." The basic geometric figures are the point, the line, and the plane. These figures are theoretical and cannot be formally defined. A **point** is represented by a dot and is named by a capital letter. A **line** is an infinite set of points and is named by a small letter or by two points on the line. A **line segment** is part of a line between two points called endpoints. A **plane** is a flat surface. See Fig. 9-1.

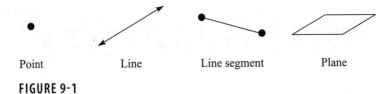

Point   Line   Line segment   Plane

FIGURE 9-1

Points and line segments are used to make geometric figures. The geometric figures presented in this chapter are the triangle, the square, the rectangle, the parallelogram, the trapezoid, and the circle. See Fig. 9-2.

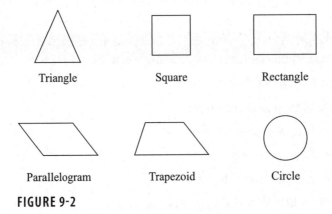

Triangle   Square   Rectangle

Parallelogram   Trapezoid   Circle

FIGURE 9-2

A **triangle** is a geometric figure with three sides. A **rectangle** is a geometric figure with four sides and four 90° angles. The opposite sides are equal in length and are parallel. A **square** is a rectangle in which all sides are of the same length. A **parallelogram** has four sides with two pairs of parallel sides. A **trapezoid** has four sides, two of which are parallel. A **circle** is a geometric figure such that all the points are the same distance from a point called its center. The center is not part of the circle. A line segment passing through the center of a circle and with its endpoints on the circle is called a **diameter**. A line segment from the center of a circle to the circle is called a **radius**. See Fig. 9-3. The radius is one-half the length of the diameter. For example, if the diameter of a circle is 10 inches, the radius is 5 inches.

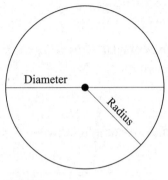

FIGURE 9-3

Perimeter

The distance around the outside of a geometric figure is called the **perimeter** of the figure. The perimeter is found by adding the measures of all sides of the geometric figure. For some geometric figures, there are special formulas that are used to find their perimeters.

The perimeter of a triangle is found by adding the lengths of the three sides: $P = a + b + c$.

EXAMPLE

Find the perimeter of the triangle shown in Fig. 9-4.

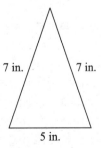

FIGURE 9-4

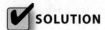 **SOLUTION**

$$P = a + b + c$$
$$= 7 \text{ inches} + 7 \text{ inches} + 5 \text{ inches}$$
$$= 19 \text{ inches}$$

The perimeter of a rectangle can be found by using the formula P = 2l + 2w, where l is the length and w is the width.

EXAMPLE

Find the perimeter of the rectangle shown in Fig. 9-5.

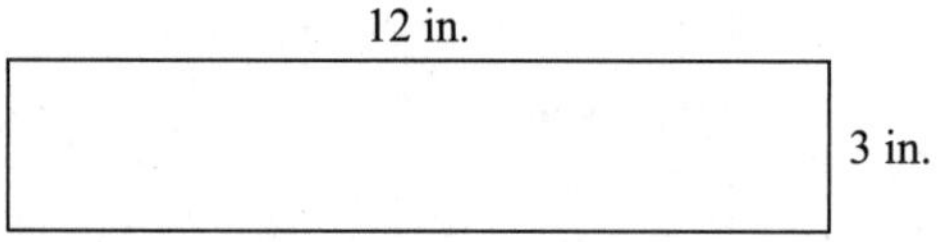

12 in.

3 in.

FIGURE 9-5

 **SOLUTION**

$$P = 2l + 2w$$
$$= 2 \cdot 12 \text{ inches} + 2 \cdot 3 \text{ inches}$$
$$= 24 + 6$$
$$= 30 \text{ inches}$$

The perimeter of a square is found by using the formula P = 4s, where s is the length of a side.

 EXAMPLE

Find the perimeter of the square shown in Fig. 9-6.

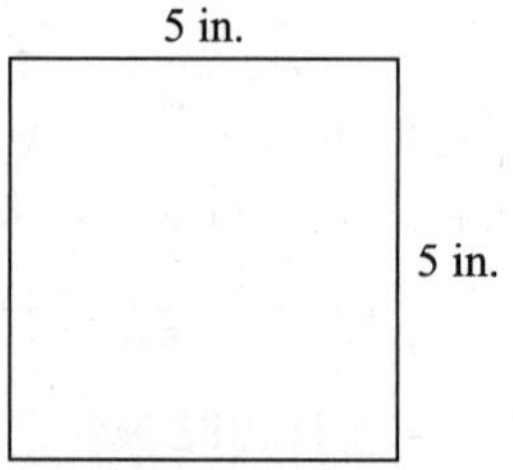

5 in.

5 in.

FIGURE 9-6

 **SOLUTION**

$$P = 4s$$
$$= 4 \cdot 5 \text{ inches}$$
$$= 20 \text{ inches}$$

The perimeter of a circle is called the **circumference** of a circle. To find the circumference of a circle, it is necessary to use either the radius (r) or the diameter (d) of the circle.

The radius of a circle is equal to one-half of its diameter. That is, $r = \frac{1}{2}d$ or $\frac{d}{2}$. The diameter of a circle is twice as large as its radius. That is, $d = 2r$.

 **EXAMPLE**

Find the radius of a circle if its diameter is 12 inches.

 **SOLUTION**

$$r = \frac{1}{2}d$$
$$= \frac{1}{2} \cdot 12 \text{ inches}$$
$$= 6 \text{ inches}$$

EXAMPLE

Find the diameter of a circle when its radius is 16 inches.

SOLUTION

$$d = 2r$$
$$= 2 \cdot 16 \text{ inches}$$
$$= 32 \text{ inches}$$

In order to find the circumference of a circle, a special number is used. That number is called pi, and its symbol is π. $\pi \approx 3.141592654\ldots$. For our purposes, we use $\pi \approx 3.14$. Pi is the number of times that a diameter of a circle will fit around the circle.

The circumference of a circle then is $C = \pi d$ or $C = 2\pi r$ where d is the diameter of the circle and r is the radius of the circle.

EXAMPLE

Find the circumference of a circle whose diameter is 9 inches. Use $\pi = 3.14$.

SOLUTION

$$C = \pi d$$
$$= 3.14 \cdot 9 \text{ inches}$$
$$= 28.26 \text{ inches}$$

EXAMPLE

Find the circumference of a circle whose radius is 8 feet. Use $\pi = 3.14$.

SOLUTION

$$C = 2\pi r$$
$$= 2 \cdot 3.14 \cdot 8 \text{ feet}$$
$$= 50.24 \text{ feet}$$

MATH NOTE *Sometimes $\frac{22}{7}$ is used as an approximate value for π.*

Practice

1. Find the perimeter of a triangle whose sides are 7 feet, 8 feet, and 10 feet.
2. Find the perimeter of a square whose side is 6 yards.
3. Find the perimeter of a rectangle whose length is 32 inches and whose width is 21 inches.
4. Find the circumference of a circle whose diameter is 14 inches. Use $\pi = 3.14$.
5. Find the circumference of a circle whose radius is 36 feet. Use $\pi = 3.14$.

Answers

1. 25 feet
2. 24 yards
3. 106 inches
4. 43.96 inches
5. 226.08 feet

Area

The **area** of a geometric figure is the number of square units contained in its surface. For example, the area of the 3-inch by 2-inch rectangle is 6 square inches as shown in Fig. 9-7.

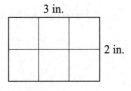

3 in.

2 in.

FIGURE 9-7

Area is measured in square units. A square inch is a square whose sides measure 1 inch. A square foot is a square whose sides measure 1 foot, etc. Square units are abbreviated using two for the exponent.

$$1 \text{ square inch} = 1 \text{ in.}^2$$

$$1 \text{ square foot} = 1 \text{ ft}^2$$

$$1 \text{ square yard} = 1 \text{ yd}^2$$

$$1 \text{ square mile} = 1 \text{ mi}^2$$

The area of a rectangle is found by using the formula A = lw, where l is the length and w is the width.

EXAMPLE

Find the area of the rectangle whose length is 5 feet and whose width is 3 feet (Fig. 9-8).

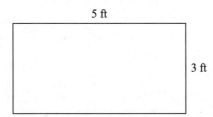

5 ft

3 ft

FIGURE 9-8

SOLUTION

$$A = lw$$
$$= 5 \text{ feet} \times 3 \text{ feet}$$
$$= 15 \text{ ft}^2$$

The area of a square can be found by using the formula $A = s^2$, where s is the length of its side.

EXAMPLE

Find the area of the square shown in Fig. 9-9.

9 in.

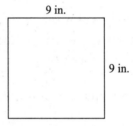

9 in.

FIGURE 9-9

SOLUTION

$$A = s^2$$
$$= (9 \text{ in.})^2$$
$$= 81 \text{ in.}^2$$

To find the area of a triangle, you need to know the measure of its *altitude* or *height*. The **altitude** or **height** of a triangle is the measure of a perpendicular line from its highest point to its base (the side opposite its highest point). See Fig. 9-10.

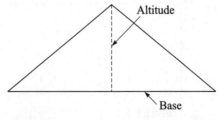

Altitude

Base

FIGURE 9-10

The area of a triangle can be found by using the formula $A = \frac{1}{2}bh$, where b is the base and h is the height or altitude.

EXAMPLE

Find the area of the triangle shown in Fig. 9-11.

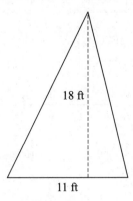

18 ft

11 ft

FIGURE 9-11

SOLUTION

$$A = \frac{1}{2}bh$$

$$= \frac{1}{2} \cdot 18\,\text{ft} \cdot 11\,\text{ft}$$

$$= 99\ \text{ft}^2$$

The area of a parallelogram can be found by using the formula $A = bh$, where b is the base and h is the height or altitude.

EXAMPLE

Find the area of the parallelogram shown in Fig. 9-12.

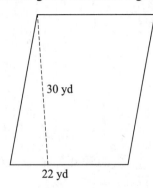

30 yd

22 yd

FIGURE 9-12

 SOLUTION

$$A = bh$$
$$= 22 \text{ yd} \cdot 30 \text{ yd}$$
$$= 660 \text{ yd}^2$$

A trapezoid has a height and two bases—a lower base, b_1, and an upper base, b_2. See Fig. 9-13.

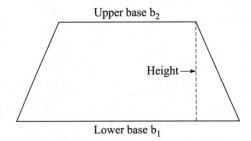

Upper base b_2

Height →

Lower base b_1

FIGURE 9-13

The area of a trapezoid can be found by using the formula $A = \dfrac{1}{2}h(b_1 + b_2)$.

EXAMPLE
Find the area of the trapezoid shown in Fig. 9-14.

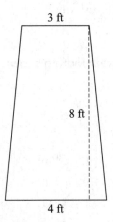

3 ft

8 ft

4 ft

FIGURE 9-14

 **SOLUTION**

$$A = \frac{1}{2}h(b_1 + b_2)$$

$$= \frac{1}{2} \cdot 8\,ft \cdot (4\,ft + 3\,ft)$$

$$= \frac{1}{2} \cdot 8\,ft \cdot 7\,ft$$

$$= 28\,ft^2$$

The area of a circle can be found by using the formula $A = \pi r^2$, where $\pi = 3.14$ and r is the radius.

 EXAMPLE

Find the area of the circle shown in Fig. 9-15. Use $\pi = 3.14$.

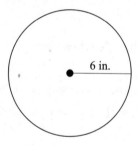

6 in.

FIGURE 9-15

 **SOLUTION**

$$A = \pi r^2$$

$$= 3.14 \cdot (6\,in.)^2$$

$$= 3.14 \cdot 36\,in.^2$$

$$= 113.04\,in.^2$$

The following suggestions will help you to convert between units. To change:

square feet to square inches, multiply by 144.

square inches to square feet, divide by 144.

square yards to square feet, multiply by 9.

square feet to square yards, divide by 9.

EXAMPLE

Change 2,160 square inches to square feet.

SOLUTION

$$2{,}160 \text{ in.}^2 \div 144 = 15 \text{ ft}^2$$

EXAMPLE

Change 12 square yards to square feet.

SOLUTION

$$12 \text{ yd}^2 \times 9 = 108 \text{ ft}^2$$

Practice

1. Find the area of a rectangle whose length is 112 yards and whose width is 63 yards.
2. Find the area of a square whose side is 47 inches.
3. Find the area of a triangle whose base is 22 inches and whose height is 7 inches.
4. Find the area of a parallelogram whose base is 18 inches and whose height is 12 inches.
5. Find the area of a trapezoid whose bases are 30 feet and 42 feet and whose height is 16 feet.
6. Find the area of a circle whose radius is 16 inches. Use $\pi = 3.14$.
7. Find the area of a circle whose diameter is 24 feet. Use $\pi = 3.14$.
8. Change 15 square feet to square inches.
9. Change 45 square feet to square yards.
10. Change 3 square yards to square inches.

Answers

1. 7,056 yd^2
2. 2,209 in.^{2}
3. 77 in.^{2}

4. 216 in.^{2}

5. 576 ft^2

6. 803.84 in.^{2}

7. 452.16 ft^2

8. 2,160 in.^{2}

9. 5 yd^2

10. 3,888 in.^{2}

Volume

The volume of a geometric figure is a measure of its capacity. Volume is measured in cubic units. Cubic units are abbreviated using 3 for the exponent.

$$1 \text{ cubic inch} = 1 \text{ in.}^3$$
$$1 \text{ cubic foot} = 1 \text{ ft}^3$$
$$1 \text{ cubic yard} = 1 \text{ yd}^3$$

The basic geometric solids are the rectangular solid, the cube, the cylinder, the sphere, the right circular cone, and the pyramid. These figures are shown in Fig. 9-16.

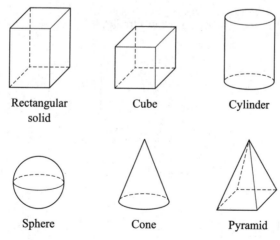

Rectangular   Cube   Cylinder
solid

Sphere   Cone   Pyramid

FIGURE 9-16

The volume of a rectangular solid can be found by using the formula $V = lwh$, where l = the length, w = the width, and h = the height.

**EXAMPLE**

Find the volume of the rectangular solid shown in Fig. 9-17.

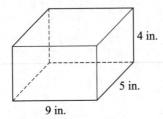

4 in.

5 in.

9 in.

FIGURE 9-17

SOLUTION

$$V = lwh$$
$$= 9 \text{ in.} \cdot 5 \text{ in.} \cdot 4 \text{ in.}$$
$$= 180 \text{ in.}^3$$

The volume of a cube can be found by using the formula $V = s^3$, where s is the length of the side.

**EXAMPLE**

Find the volume of the cube shown in Fig. 9-18.

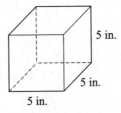

5 in.

5 in.

5 in.

FIGURE 9-18

**SOLUTION**

$$V = s^3$$
$$= (5 \text{ in.})^3$$
$$= 125 \text{ in.}^3$$

The volume of a cylinder can be found by using the formula $V = \pi r^2 h$, where r is the radius of the base and h is the height.

 EXAMPLE

Find the volume of the cylinder shown in Fig. 9-19. Use $\pi = 3.14$.

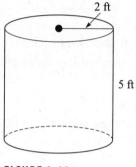

2 ft

5 ft

FIGURE 9-19

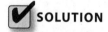 **SOLUTION**

$$V = \pi r^2 h$$
$$= 3.14 \cdot (2 \text{ ft})^2 \cdot 5 \text{ ft}$$
$$= 62.8 \text{ ft}^3$$

The volume of a sphere can be found by using the formula $V = \dfrac{4}{3}\pi r^3$, where r is the radius of the sphere.

 EXAMPLE

Find the volume of the sphere shown in Fig. 9-20. Use $\pi = 3.14$.

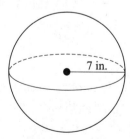

7 in.

FIGURE 9-20

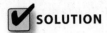 SOLUTION

$$V = \frac{4}{3}\pi r^3$$

$$= \frac{4}{3} \cdot 3.14 \cdot (7 \text{ in.})^3$$

$$= 1{,}436.03 \text{ in.}^3 \text{ (rounded)}$$

The volume of a right circular cone can be found by using the formula $V = \frac{1}{3}\pi r^2 h$, where r is the radius of the base and h is the height of the cone.

 EXAMPLE

Find the volume of the cone shown in Fig. 9-21. Use $\pi = 3.14$.

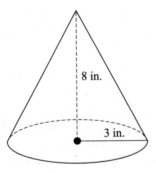

FIGURE 9-21

 SOLUTION

$$V = \frac{1}{3}\pi r^2 h$$

$$= \frac{1}{3} \cdot 3.14 \cdot (3\text{ in.})^2 \cdot 8\text{ in.}$$

$$= 75.36 \text{ in.}^3$$

The volume of a pyramid can be found by using the formula $V = \frac{1}{3}Bh$, where B is the area of the base and h is the height of the pyramid. If the base is a square, use $B = s^2$. If the base is a rectangle, use $B = lw$.

EXAMPLE

Find the volume of the pyramid with a square base shown in Fig. 9-22.

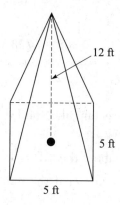

12 ft

5 ft

5 ft

FIGURE 9-22

 SOLUTION

In this case, the base is a square, so the area of the base is $B = s^2$.

$$V = \frac{1}{3}B \cdot h$$

$$= \frac{1}{3}(5\,\text{ft})^2 \cdot 12\,\text{ft}$$

$$= 100\,\text{ft}^3$$

Sometimes it is necessary to convert from cubic yards to cubic feet, cubic inches to cubic feet, etc. The following information will help you to do this.
   To change:

   cubic feet to cubic inches, multiply by 1,728
   cubic inches to cubic feet, divide by 1,728
   cubic yards to cubic feet, multiply by 27
   cubic feet to cubic yards, divide by 27

 **EXAMPLE**

Change 14 cubic yards to cubic feet.

 SOLUTION

$$14 \times 27 = 378\,\text{ft}^3$$

**EXAMPLE**

Change 10,368 cubic inches to cubic feet.

SOLUTION

$$10{,}368 \div 1{,}728 = 6 \text{ ft}^3$$

Practice

1. Find the volume of a rectangular solid if it is 18 inches long, 15 inches wide, and 13 inches high.

2. Find the volume of a cube if the side is 14 feet.

3. Find the volume of a cylinder if the radius is 5 inches and its height is 9 inches. Use $\pi = 3.14$.

4. Find the volume of a sphere if its radius is 6 inches. Use $\pi = 3.14$.

5. Find the volume of a cone if its radius is 5 feet and its height is 12 feet. Use $\pi = 3.14$.

6. Find the volume of a pyramid if its height is 6 yards and its base is a rectangle whose length is 5 yards and whose width is 3.5 yards.

Answers

1. 3,510 in.^{3}

2. 2,744 ft^3

3. 706.5 in.^{3}

4. 904.32 in.^{3}

5. 314 ft^3

6. 35 yd^3

Word Problems

Word problems in geometry can be solved by the following procedure.

> *Step 1: Read the problem.*
>
> *Step 2: Decide whether you are being asked to find the perimeter, area, or volume.*
>
> *Step 3: Select the correct formula.*
>
> *Step 4: Substitute in the formula and evaluate.*

EXAMPLE

How many inches of wood framing are needed for a window frame that measures 72 inches by 48 inches?

 SOLUTION

Since the frame goes along the edge of the window, it is necessary to find the perimeter of a rectangle 72 inches by 48 inches.

$$P = 2l + 2w$$
$$= 2 \cdot 72 \text{ in.} + 2 \cdot 48 \text{ in.}$$
$$= 240 \text{ in.}$$

 EXAMPLE

A silo is 24 feet high and has a diameter of 10 feet. How many cubic feet of corn will it hold? Use $\pi = 3.14$.

 SOLUTION

Find the volume of a cylinder.

$$V = \pi r^2 h \qquad r = 10 \text{ ft} \div 2 = 5 \text{ ft}$$
$$V = 3.14 \cdot (5 \text{ ft})^2 \cdot 24 \text{ ft}$$
$$V = 1,884 \text{ ft}^3$$

Practice

1. A circular garden pool has a diameter of 36 inches and is 8 inches deep. How many cubic inches of water will it hold? Use $\pi = 3.14$.

2. How many cubic yards of sand are in a conical pile of sand that is 2 feet high and has a diameter of $1\frac{1}{2}$ feet? Use $\pi = 3.14$.

3. Find the weight of a steel bar if it is 5 feet long, 2 inches thick, and 4 inches wide. One cubic foot of steel weighs 480 pounds.

4. Find the volume of a cubic bin if it is 4 feet on a side.

5. Find the volume of a spherical globe that has a diameter of 12 inches. Use $\pi = 3.14$.

Answers

1. 8,138.88 in^3

2. 1.1775 ft^3

3. $133\frac{1}{3}$ pounds

4. 64 ft^3

5. 904.32 in^3

Square Roots

In Chap. 2 you learned how to square a number. For example, $7^2 = 7 \times 7 = 49$ and $3^2 = 3 \times 3 = 9$. The opposite of squaring a number is taking the square root of a number. The radical sign ($\sqrt{\ }$) is used to indicate the square root of a number, so $\sqrt{16} = 4$ because $4^2 = 16$ and $\sqrt{25} = 5$ because $5^2 = 25$.

Numbers such as 1, 4, 9, 16, 25, 36, 49, etc., are called perfect squares because their square roots are rational numbers.

The square roots of other numbers such as $\sqrt{2}, \sqrt{3}, \sqrt{5}$, etc., are called **irrational** numbers because their square roots are infinite, non-repeating decimals. For example, $\sqrt{2} = 1.414213562 \ldots$ and $\sqrt{3} = 1.732050808 \ldots$. In other words, the decimal value of $\sqrt{2}$ cannot be found exactly.

Still Struggling

The easiest way to find the square root of a number is to use a calculator. However, the square root of a perfect square can be found by guessing and then squaring the answer to see if it correct. For example, to find $\sqrt{196}$, you could guess it is 12. Then square 12. $12^2 = 144$. This is too small. Try 13. $13^2 = 169$. This is still too small, so try 14. $14^2 = 196$. Hence, $\sqrt{196} = 14$.

The set of numbers which consists of the rational numbers and the irrational numbers is called the set of **real numbers**.

The Pythagorean Theorem

An important mathematical principle is called the Pythagorean Theorem. This principle uses right triangles. A **right triangle** is a triangle which has one right or 90° angle. The side opposite the 90° angle is called the **hypotenuse**.

The Pythagorean Theorem states that for any right triangle, $c^2 = a^2 + b^2$, where c is the length of the hypotenuse and a and b are the lengths of its sides. See Fig. 9-23.

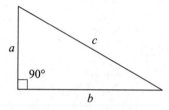

FIGURE 9-23

If you need to find the hypotenuse of a right triangle, use $c = \sqrt{a^2 + b^2}$. If you need to find the length of one side of a right triangle, use $a = \sqrt{c^2 - b^2}$ or $b = \sqrt{c^2 - a^2}$.

 EXAMPLE

Find the length of the hypotenuse if the lengths of its sides are 10 inches and 24 inches.

 SOLUTION

$$c = \sqrt{a^2 + b^2}$$
$$= \sqrt{10^2 + 24^2}$$
$$= \sqrt{100 + 576}$$
$$= \sqrt{676}$$
$$= 26 \text{ inches}$$

 EXAMPLE

An airplane flies 150 miles due east, then makes a left turn and flies 360 miles north. How far is the plane from the starting point?

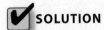**SOLUTION**

The right triangle indicating the path of the plane is shown in Fig. 9-24.

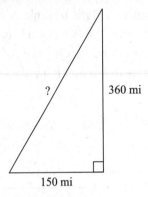

360 mi

?

150 mi

FIGURE 9-24

Then

$$c = \sqrt{a^2 + b^2}$$
$$= \sqrt{150^2 + 360^2}$$
$$= \sqrt{22,500 + 129,600}$$
$$= \sqrt{152,100}$$
$$= 390 \text{ miles}$$

Practice

1. Find the length of the hypotenuse of a right triangle whose sides are 60 inches and 25 inches.

2. Find the length of a side of a right triangle whose hypotenuse is 35 yards and whose other side is 28 yards.

3. Find the distance that you would walk up a staircase if it is 45 feet high and has a base of 28 feet.

4. Find the perimeter of a flower garden if its shape is a right triangle with a hypotenuse of 13 feet and a side of 5 feet.

5. Find the hypotenuse of a sail whose shape is a right triangle with a base of 48 feet and whose height is 55 feet.

Answers

1. $\sqrt{4,225} = 65$ inches
2. $\sqrt{441} = 21$ yards
3. $\sqrt{2,809} = 53$ feet
4. 30 feet
5. $\sqrt{5,329} = 73$ feet

In this chapter, you learned how to find the perimeter, area, and volume of the basic geometric figures. The Pythagorean Theorem and its applications were shown.

QUIZ

1. Find the perimeter of a rectangle whose length is 21 feet and whose width is 15 feet.
   A. 36 ft
   B. 50 ft
   C. 72 ft
   D. 315 ft

2. Find the circumference of a circle whose diameter is 19 inches. Use $\pi = 3.14$.
   A. 59.66 in.
   B. 29.83 in.
   C. 1,133.54 in.
   D. 119.32 in.

3. Find the perimeter of a square whose side is 33 inches.
   A. 66 in.
   B. 132 in.
   C. 99 in.
   D. 165 in.

4. Find the perimeter of a triangle whose sides are 8 yards, 6 yards, and 11 yards.
   A. 12.5 yd
   B. 14 yd
   C. 48 yd
   D. 25 yd

5. If the radius of a circle is 9.8 inches, find the diameter.
   A. 19.6 in.
   B. 4.9 in.
   C. 9.8 in.
   D. 96.04 in.

6. Find the area of a circle whose radius is 7 inches. Use $\pi = 3.14$.
   A. 21.98 in.^{2}
   B. 43.96 in.^{2}
   C. 153.86 in.^{2}
   D. 615.44 in.^{2}

7. Find the area of a square whose side is 8.3 feet.
   A. 33.2 ft^2
   B. 41.5 ft^2
   C. 68.89 ft^2
   D. 16.6 ft^2

8. **Find the area of a trapezoid whose height is 15 inches and whose bases are 6 inches and 16 inches.**

    A. 93 in.^{2}
    B. 165 in.^{2}
    C. 330 in.^{2}
    D. 168 in.^{2}

9. **Find the area of a rectangle whose length is 9.5 inches and whose width is 6.2 inches.**

    A. 15.7 in.^{2}
    B. 31.4 in.^{2}
    C. 29.45 in.^{2}
    D. 58.9 in.^{2}

10. **Find the area of a triangle whose base is 56 yards and whose height is 45 yards.**

    A. 1,260 yd^2
    B. 2,520 yd^2
    C. 202 yd^2
    D. 303 yd^2

11. **Find the area of a parallelogram whose base is 12.6 inches and whose height is 5.3 inches.**

    A. 17.9 in.^{2}
    B. 35.8 in.^{2}
    C. 33.39 in.^{2}
    D. 66.78 in.^{2}

12. **How many square inches are there in 17 square feet?**

    A. 51 in.^{2}
    B. 2,448 in.^{2}
    C. 289 in.^{2}
    D. 68 in.^{2}

13. **How many square yards are there in 1,287 square feet?**

    A. 429 yd^2
    B. 540 yd^2
    C. 143 yd^2
    D. 107.25 yd^2

14. **Find the volume of a sphere whose radius is 15 inches. Use $\pi = 3.14$.**

    A. 942 in.^{3}
    B. 14,130 in.^{3}
    C. 47.1 in.^{3}
    D. 529.875 in.^{3}

15. Find the volume of a cube whose side is 6.1 inches in length.
    A. 18.3 in.^{3}
    B. 37.21 in.^{3}
    C. 24.4 in.^{3}
    D. 226.981 in.^{3}

16. Find the volume of a cylinder whose base has a radius of 6 inches and whose height is 12 inches. Use $\pi = 3.14$.
    A. 1,356.48 in.^{3}
    B. 339.12 in.^{3}
    C. 113.04 in.^{3}
    D. 452.16 in.^{3}

17. Find the volume of a rectangular solid whose length is 17 inches, whose width is 8 inches, and whose length is 3 inches.
    A. 28 in.^{3}
    B. 59 in.^{3}
    C. 408 in.^{3}
    D. 160 in.^{3}

18. Find the volume of a cone whose base has a diameter of 12 feet and whose height is 9 feet. Use $\pi = 3.14$.
    A. 339.12 ft^3
    B. 405.06 ft^3
    C. 36 ft^3
    D. 1,356.48 ft^3

19. Find the volume of a pyramid whose height is 4 feet and whose base is a square whose side is 6 feet.
    A. 24 ft^3
    B. 48 ft^3
    C. 32 ft^3
    D. 8 ft^3

20. Find the length of the hypotenuse of a right triangle whose sides are 180 inches and 112 inches.
    A. 142 in.
    B. 292 in.
    C. 472 in.
    D. 212 in.

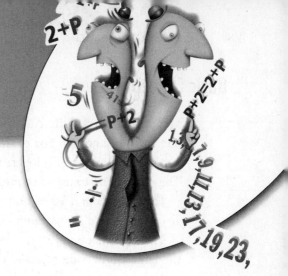

chapter **10**

Measurement

This chapter explains how to convert basic measurements from one unit to another. For example, you may need to cover an item with material and you have measured the dimensions of the item in square feet. When you go to purchase the material, you find that it is sold in square yards, so you must convert square feet to square yards. This chapter explains measures of length, weight, capacity, and time. There are many applications of measurement that you will come across in your lifetime.

CHAPTER OBJECTIVES

In this chapter, you will learn how to

- Convert measures of length into different units
- Convert measures of weight into different units
- Convert measures of capacity into different units
- Convert measures of time into different units
- Solve word problems involving different measures

Basic Concepts

Many times in real-life situations, you must convert measurement units from one form to another. For example, you may need to change feet to yards, or quarts to gallons, etc. There are several mathematical methods that can be used. The easiest one to use is to multiply or divide.

To change from a larger unit to a smaller unit, **multiply** *by the conversion unit. To change from a smaller unit to a larger one,* **divide** *by the conversion unit.*

The conversion units are arranged in descending order in each table. That is, the largest unit is on top, and the smallest unit is on the bottom. This chapter includes conversions for length, weight, capacity, and time. Only the most common units of measurement are explained in this chapter. Units such as rods, nautical miles, fathoms, and barrels have been omitted.

Measures of Length

Length is commonly measured in inches, feet, yards, and miles. (Table 10.1)

TABLE 10-1 Conversion Factors for Length

1 mile (mi) = 1,760 yards (yd) = 5,280 feet (ft)

1 yard (yd) = 3 feet (ft)

1 foot (ft) = 12 inches (in.)

EXAMPLE
Change 7 yards to feet.

SOLUTION
Since we are going from large to small, and the conversion factor is 1 yd = 3 ft, we multiply by 3 ft.

$$7 \text{ yd} \times 3 \text{ ft} = 21 \text{ ft}$$

EXAMPLE

Change 672 inches to feet.

SOLUTION

Since we are going from small to large, and the conversion factor is 1 ft = 12 in., we divide:

$$672 \text{ in.} \div 12 \text{ in.} = 56 \text{ ft}$$

Sometimes it is necessary to use an operation more than once to arrive at the solution.

EXAMPLE

Change 89,760 feet to miles.

SOLUTION

$$89,760 \text{ ft} \div 3 \text{ ft} = 29,920 \text{ yd}$$

$$29,920 \text{ yds} \div 1,760 \text{ yd} = 17 \text{ miles}$$

MATH NOTE *Since you know the conversion factor (1 mile = 5,280 feet), it is only necessary to divide once: 89,760 ft ÷ 5,280 ft = 17 mi*

Another type of conversion has mixed units.

EXAMPLE

Change 15 feet 9 inches to inches.

SOLUTION

First change 15 feet to inches and then add 9 inches to the answer.

$$15 \text{ ft} \times 12 \text{ in.} = 180 \text{ in.}$$

$$180 \text{ in.} + 9 \text{ in.} = 189 \text{ in.}$$

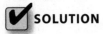**EXAMPLE**

Change 47 feet to yards and write the remainder in feet.

SOLUTION

$$47 \text{ ft} \div 3 \text{ ft} \qquad 3\overline{)47} \atop \begin{array}{r} 15 \\ \underline{3} \\ 17 \\ \underline{15} \\ 2 \end{array}$$

Hence, the answer is 15 yd 2 ft.

MATH NOTE *The answer in the previous problem could also be written as* $15\frac{2}{3}$ *yards.*

**EXAMPLE**

Change 9 yards 16 inches to yards.

SOLUTION

$$9 \text{ yd } 16 \text{ in.} = 9 \text{ yd} + 16 \text{ in.}$$
$$= 9 \text{ yd} + \frac{16}{36} \text{ in.} \qquad \text{Change inches to yards}$$
$$= 9\frac{16}{36} \text{ yd}$$
$$= 9\frac{4}{9} \text{ yd}$$

Practice
Change:

1. 19 ft to inches

2. 216 ft to yards

3. 7 ft 10 in. to inches

4. 12 yd to inches

5. 14,080 yd to miles

6. 9 mi to feet

7. 73,920 ft to miles

8. 7,000 yd to miles (Write the remainder in yards.)

9. 23 ft 8 in. to feet

10. 16 yd 7 ft 4 in. to inches

Answers

1. 228 in.

2. 72 yd

3. 94 in.

4. 432 in.

5. 8 mi

6. 47,520 ft

7. 14 mi

8. 3 mi 1720 yd

9. $23\frac{2}{3}$ ft

10. 664 in.

Measures of Weight

Weight is commonly measured in ounces, pounds, and tons. Use the conversion factors in Table 10.2 and follow the rules stated in the Basic Concepts section.

TABLE 10-2 Conversion Factors for Weight
1 ton (T) = 2,000 pounds (lb)
1 pound (lb) = 16 ounces (oz)

EXAMPLE

Change 6 pounds to ounces.

SOLUTION

$$6 \text{ lb} \times 16 \text{ oz} = 96 \text{ oz}$$

EXAMPLE

Change 14,000 pounds to tons.

SOLUTION

$$14{,}000 \text{ lb} \div 2{,}000 \text{ T} = 7 \text{ T}$$

EXAMPLE

Change 3 tons to ounces.

SOLUTION

$$3 \text{ T} \times 2{,}000 \text{ lb} = 6{,}000 \text{ lb}$$

$$6{,}000 \text{ lb} \times 16 \text{ oz} = 96{,}000 \text{ oz}$$

EXAMPLE

Change 19 pounds 6 ounces to ounces.

SOLUTION

$$19 \text{ lb} \times 16 \text{ oz} = 304 \text{ oz}$$

$$304 \text{ oz} + 6 \text{ oz} = 310 \text{ oz}$$

EXAMPLE

Change 84 ounces to pounds and write the remainder in ounces.

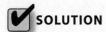 **SOLUTION**

$$84 \text{ oz} \div 16 \text{ oz} \qquad 16\overline{\smash)84} \\ \underline{80} \\ 4$$

Hence, the answer is 5 lb 4 oz.

Practice

Change:

1. 17 lb to ounces

2. 38,000 lb to tons

3. 4 lb 9 oz to ounces

4. 62 oz to pounds (Write the remainder in ounces.)

5. 6.4 T to ounces

6. 7,000 oz to pounds

7. 3 oz to pounds

8. 5 T 400 lb to pounds

9. 2,700,000 oz to tons

10. $8\frac{1}{2}$ lb to ounces

Answers

1. 272 oz

2. 19 T

3. 73 oz

4. 3 lb 14 oz

5. 204,800 oz

6. 437.5 lb

7. 0.1875 lb

8. 10,400 lb

9. 84.375 T

10. 136 oz

Measures of Capacity

Liquids are commonly measured in ounces, pints, quarts, and gallons. (Table 10.3)

TABLE 10-3	Conversion Factors for Capacity (Liquid)
1 gallon (gal) = 4 quarts (qt)	
1 quart (qt) = 2 pints (pt)	
1 pint (pt) = 16 ounces (oz)	

Still Struggling

You should be careful to distinguish between a liquid ounce, which is a measure of capacity, and an ounce, which is a measure of weight. Both measures are called ounces, but they are different. Liquid ounces are often called fluid ounces and abbreviated as fl. oz.

EXAMPLE
Change 9 gallons to quarts.

 **SOLUTION**

$$9 \text{ gal} \times 4 \text{ qt} = 36 \text{ qt}$$

EXAMPLE
Change 19 pints to quarts.

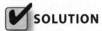 **SOLUTION**

$$19 \text{ pt} \div 2 \text{ pt} = 9.5 \text{ qt}$$

EXAMPLE

Change 4 pints 12 ounces to pints.

✔ **SOLUTION**

$$4 \text{ pt } 12 \text{ oz} = 4 \text{ pt} + 12 \text{ oz}$$
$$= 4 \text{ pt} + \frac{12}{16} \text{ pt}$$
$$= 4\frac{12}{16} \text{ pt}$$
$$= 4\frac{3}{4} \text{ pt}$$

 **EXAMPLE**

Change 104 pints to gallons.

✔ **SOLUTION**

$$104 \text{ pt} \div 2 \text{ pt} = 52 \text{ qt}$$
$$52 \text{ qt} \div 4 \text{ qt} = 13 \text{ gal}$$

Practice

Change:

1. 9 pt to ounces
2. 280 oz to pints
3. 6 pt 8 oz to ounces
4. 21 qt to pints
5. 10 gal to ounces
6. 50 oz to pints (Write the remainder in ounces.)
7. 7 qt 1 pt to quarts
8. 61 pt to gallons (Write the remainder in quarts and pints.)
9. 5.5 gal to pints
10. 45 oz to quarts

Answers

1. 144 oz
2. 17.5 pt

3. 104 oz

4. 42 pt

5. 1,280 oz

6. 3 pt 2 oz

7. 7.5 qt

8. 7 gal 2 qt 1 pt

9. 44 pt

10. $1\frac{13}{32}$ qt

Measures of Time

Time is measured in years, weeks, days, hours, minutes, and seconds. (Table 10.4)

TABLE 10-4 Conversion Factors for Time
1 year (yr) = 12 months (mo)
= 52 weeks (wk)
= 365 days (da)*
1 week (wk) = 7 days (da)
1 day (da) = 24 hours (hr)
1 hour (hr) = 60 minutes (min)
1 minute (min) = 60 seconds (sec)

*Leap years will not be used.

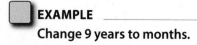

EXAMPLE

Change 9 years to months.

SOLUTION

$$9 \text{ yr} \times 12 \text{ mo} = 108 \text{ mo}$$

EXAMPLE

Change 2,920 days to years.

SOLUTION

$$2{,}920 \text{ da} \div 365 \text{ da} = 8 \text{ yr}$$

EXAMPLE

Change 64 days to weeks. (Write the remainder in days.)

SOLUTION

$$\begin{array}{r} 9 \\ 7{\overline{)64}} \\ \underline{63} \\ 1 \end{array}$$

64 days = 9 wk 1 da

Practice

Change:

1. 1,860 min to hours
2. 32 wk to days
3. 7 hr 42 min to minutes
4. 86 mo to years (Write the remainder in months.)
5. 45 min to hours
6. 31 hr 52 min to seconds
7. 6.5 yr to weeks
8. 5 hr 15 min to hours
9. 17 min 32 sec to seconds
10. 6 yr to days

Answers

1. 31 hr
2. 224 da

3. 462 min

4. 7 yr 2 mo

5. $\frac{3}{4}$ or 0.75 hr

6. 114,720 sec

7. 338 wk

8. 5.25 hr or $5\frac{1}{4}$ hr

9. 1,052 sec

10. 2,190 da

Word Problems

The following procedure can be used to solve word problems involving measurement:

Step 1: Read the problem and decide what you are being asked to find.

Step 2: Select the correct conversion factor or factors.

Step 3: Multiply or divide.

Step 4: Complete any additional steps necessary.

 EXAMPLE

Find the cost of 18 feet of rope if it costs $0.72 per yard.

 SOLUTION

Change 18 feet to yards: 18 ft ÷ 3 ft = 6 yd

Find the cost of 6 yards: $0.72 × 6 = $4.32

Hence, 18 feet of rope will cost $4.32.

 EXAMPLE

If a gallon of oil costs $9.95 and a quart of oil costs $2.59, how much is saved by buying the oil in gallons instead of quarts if a person needs 4 gallons of oil?

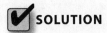 **SOLUTION**

Cost of 4 gallons of oil: $9.95 × 4 = $39.80

Change 4 gallons to quarts: 4 gal × 4 qt = 16 qt

Cost of 16 qt of oil: $2.59 × 16 = $41.44

Subtract: $41.44 – $39.80 = $1.64

Hence, a person can save $1.64 by buying the oil in gallons instead of quarts.

Practice

1. It takes 9 inches of ribbon to make a decorative identification badge. Find the cost of making 40 badges if ribbon costs $0.59 per yard?

2. If the speed of light is 186,000 miles per second, how long in hours does it take the light from the sun to reach the planet Mars, if it is 1,435,000,000 miles from the sun?

3. How many cubic feet of salt water are in a container that weighs 3 tons? One cubic foot of water weighs 64 pounds. (Ignore the weight of the container.)

4. How many 8-ounce boxes of peanuts can be filled from a 4-pound box of candy?

5. If a family uses $1\frac{1}{2}$ quarts of orange juice per week, how many gallons are used in 1 year?

Answers

1. $5.90
2. 2.14 hr (rounded)
3. 93.75 ft³
4. 8 boxes
5. 19.5 gal

This chapter presented the basic concepts of measurement which include measures of length, weight, capacity, and time. Many situations in real life require you to convert from one measurement unit to another.

QUIZ

1. **Change 16 feet to inches.**

   A. $1\frac{1}{3}$ in.
   B. 192 in.
   C. 384 in.
   D. 2,304 in.

2. **Change 84 feet to yards.**

   A. 28 yd
   B. 25 yd
   C. 225 yd
   D. 252 yd

3. **Change 34,848 feet to miles.**

   A. 2.2 mi
   B. 4.4 mi
   C. 8.8 mi
   D. 6.6 mi

4. **Change 7 yards 1 foot to inches.**

   A. 22 in.
   B. 84 in.
   C. 264 in.
   D. 328 in.

5. **Change 5,808 yards to miles.**

   A. 2.3 mi
   B. 0.23 mi
   C. 3.3 mi
   D. 0.33 mi

6. **Change 14 yards, 8 feet, 7 inches to inches.**

   A. 504 in.
   B. 607 in.
   C. 600 in.
   D. 336 in.

7. **Change 51 pounds to ounces.**

   A. 408 oz
   B. 2,753 oz
   C. 17 oz
   D. 816 oz

8. **Change 12,800 pounds to tons.**

    A. 6.4 T
    B. 640 T
    C. 64 T
    D. 6 T

9. **Change 72 ounces to pounds and write the remainder in ounces.**

    A. 3 lb 7 oz
    B. 4 lb 8 oz
    C. 4 lb 5 oz
    D. 3 lb 14 oz

10. **Change 4 tons to ounces.**

    A. 128,000 oz
    B. 8,000 oz
    C. 80,000 oz
    D. 12,800 oz

11. **Change $5\frac{3}{4}$ pounds to ounces.**

    A. 46 oz
    B. 92 oz
    C. 110 oz
    D. 192 oz

12. **Change 7,000 pounds to tons.**

    A. 14,000 T
    B. 7 T
    C. 10 T
    D. 3.5 T

13. **Change 6.6 gallons to ounces.**

    A. 64 oz
    B. 56 oz
    C. 844.8 oz
    D. 144 oz

14. **Change 16 quarts to pints.**

    A. 32 pt
    B. 16 pt
    C. 8 pt
    D. 4 pt

15. **Change 14 pints 7 ounces to ounces.**
    A. 224 oz
    B. 231 oz
    C. 112 oz
    D. 108 oz

16. **Change 63 months to years.**
    A. 5 yr
    B. 75. 6 yr
    C. 7.56 yr
    D. 5.25 yr

17. **Change 18.75 seconds to minutes.**
    A. 0.3 min
    B. 3 min
    C. 0.3125 min
    D. 31.25 min

18. **Change 74 days into weeks and write the remainder in days.**
    A. 10 wk 4 da
    B. 5 wk 1 da
    C. 8 wk 3 da
    D. 9 wk 6 da

19. **How high in miles is Pike's Peak if it is 14,110 feet high?**
    A. about 6.2 mi
    B. about 4.6 mi
    C. about 2.7 mi
    D. about 1.8 mi

20. **Find the cost per ounce of orange juice if a gallon of juice sells for $3.52.**
    A. $0.0275
    B. $0.0625
    C. $0.1450
    D. $0.0860

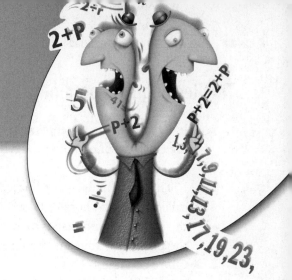

chapter **11**

Graphing

This chapter presents the basic concepts of graphing points and lines on the rectangular coordinate plane. These concepts are necessary if you plan to continue your study of mathematics. There are many applications of graphing in other areas of mathematics and science. The graphic solution to a system of linear equations is also presented in this chapter.

CHAPTER OBJECTIVES

In this chapter, you will learn how to

- Plot points
- Graph lines
- Find the x and y intercepts of a line
- Find the slope of a line
- Solve a system of two linear equations by graphing

The Rectangular Coordinate Plane

In Chap. 9 some concepts of geometry were explained. This chapter explains the concepts of combining algebra and geometry using the **rectangular coordinate plane.**

The rectangular coordinate plane uses two perpendicular number lines called **axes.** The horizontal axis is called the **x axis.** The vertical axis is called the **y axis.** The intersection of the axes is called the **origin.** The axes divide the plane into four quadrants. These quadrants are called QI, QII, QIII, and QIV. See Fig. 11-1.

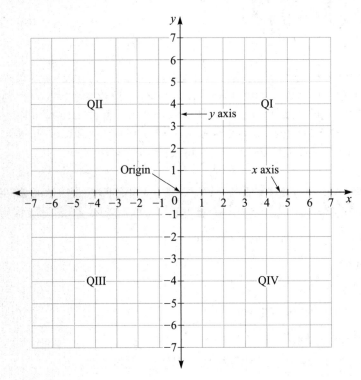

FIGURE 11-1

Plotting Points

Each point on the plane can be located by its **coordinates.** The coordinates give the horizontal and vertical distances from the y axis and x axis, respectively. The distances are called the **x coordinate (abscissa)** and the **y coordinate**

(**ordinate**), and they are written as an ordered pair (x, y). For example, a point with coordinates (2, 3) is located 2 units to the right of the y axis and 3 units above the x axis. See Fig. 11-2.

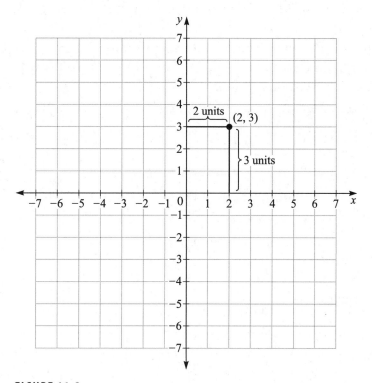

FIGURE 11-2

The point whose coordinates are (2, 3) is located in the first quadrant or QI since both coordinates are positive. The point whose coordinates are (–4, 1) is located in the second quadrant or QII since the x coordinate is negative and the y coordinate is positive. The point whose coordinates are (–3, –5) is in the third quadrant or QIII since both coordinates are negative. The point (3, –1) is located in the fourth quadrant or QIV since the x coordinate is positive and the y coordinate is negative. See Fig. 11-3.

The coordinates of the origin are (0, 0).

Any point whose y coordinate is zero is located on the x axis. For example, a point whose coordinates are (–3, 0), is located on the x axis 3 units to the left of the y axis. Any point whose x coordinate is zero is located on the y axis. For example, a point whose coordinates are (0, 4), is located on the y axis 4 units above the x axis. See Fig. 11-4.

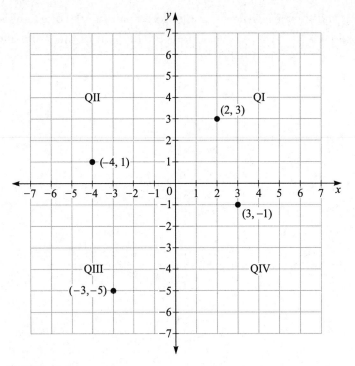

FIGURE 11-3

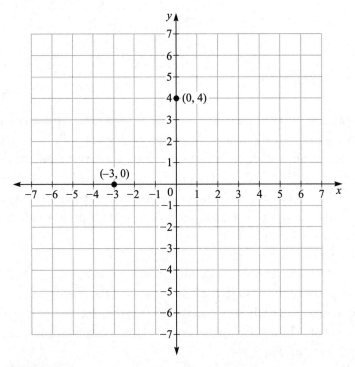

FIGURE 11-4

EXAMPLE

Give the coordinates of each point shown in Fig. 11-5.

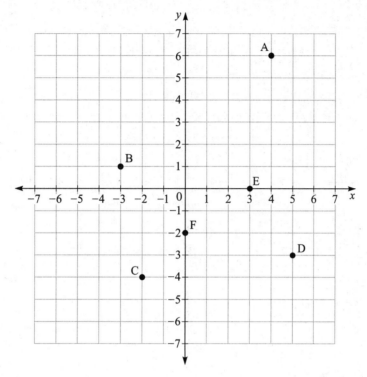

FIGURE 11-5

 **SOLUTION**

A (4, 6); B (–3, 1); C (–2, –4); D (5, –3); E (3, 0); F (0, –2)

Practice

Give the coordinates of each point shown in Fig. 11-6.

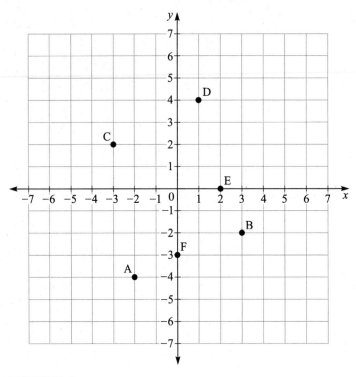

FIGURE 11-6

Answers

1. A (–2, –4)
2. B (3, –2)
3. C (–3, 2)
4. D (1, 4)
5. E (2, 0)
6. F (0, –3)

Linear Equations

An equation such as x + 2y = 6 is called a **linear equation** in two variables. A solution to the equation is a set of ordered pairs (x, y) such that when the values are substituted for the variables in the equation, the result is a closed true equation. For example, (2, 2) is a solution since

$$x + 2y = 6$$

$$2 + 2(2) = 6$$

$$2 + 4 = 6$$

$$6 = 6$$

Another solution to the equation x + 2y = 6 is (4, 1) since

$$x + 2y = 6$$

$$4 + 2(1) = 6$$

$$4 + 2 = 6$$

$$6 = 6$$

The linear equation x + 2y = 6 actually has an *infinite* number of solutions. For any value of x a corresponding value of y can be found by substituting the value for x into the equation and solving it for y. You may also get fractional or decimal answers. For example, if x = 8, then

$$x + 2y = 6$$

$$8 + 2y = 6$$

$$8 - 8 + 2y = 6 - 8$$

$$2y = -2$$

$$\frac{\overset{1}{\cancel{2}}y}{\underset{1}{\cancel{2}}} = \frac{-2}{2}$$

$$y = -1$$

When x = 8, y = −1 and the ordered pair (8, −1) is a solution, too.

EXAMPLE

Find a solution to the equation 4x – y = 8.

SOLUTION

Select any value for x, say x = 3, substitute in the equation, and then solve for y.

$$4x - y = 8$$
$$4(3) - y = 8$$
$$12 - y = 8$$
$$12 - 12 - y = 8 - 12$$
$$-y = -4$$
$$\frac{\overset{1}{\cancel{-1}}y}{\underset{1}{\cancel{-1}}} = \frac{-4}{-1}$$
$$y = 4$$

Hence (3, 4) is a solution to 4x – y = 8.

MATH NOTE *The solution can be checked by substituting both values in the equation.*

$$4x - y = 8$$
$$4(3) - 4 = 8$$
$$12 - 4 = 8$$
$$8 = 8$$

 EXAMPLE

Given the equation 3x – 4y = 26, find y when x = 2.

 SOLUTION

Substitute 2 in the equation and solve for y.

$$3x - 4y = 26$$
$$3(2) - 4y = 26$$
$$6 - 4y = 26$$
$$6 - 6 - 4y = 26 - 6$$
$$-4y = 20$$
$$\frac{\overset{1}{\cancel{-4}}y}{\underset{1}{\cancel{-4}}} = \frac{20}{-4}$$
$$y = -5$$

Hence, when x = 2, y = –5, the ordered pair (2, –5) is a solution for 3x – 4y = 26.

Practice

1. Find y when x = 5 for 2x – 4y = 2.
2. Find y when x = 1 for x + 5y = 16.
3. Find y when x = 0 for 3x + 6y = –30.
4. Find y when x = –6 for –x + 4y = 10.
5. Find y when x = –4 for 5x – 4y = 20.

Answers

1. y = 2
2. y = 3
3. y = –5
4. y = 1
5. y = –10

Graphing Lines

Equations of the form ax + by = c where a, b, and c are real numbers are called **linear equations in two variables** and their graphs are straight lines.

In order to graph a linear equation, find the coordinates of two points on the line (i.e., solutions), and then plot the points and draw the line through the two points.

 EXAMPLE

Draw the graph of x + y = 4.

 SOLUTION

Select any two values for x and find the corresponding y values.

Let x = 3	Let x = 6
x + y = 4	x + y = 4
3 + y = 4	6 + y = 4
3 – 3 + y = 4 – 3	6 – 6 + y = 4 – 6
y = 1	y = –2
(3, 1)	(6, –2)

Plot the points and draw a line through the two points. See Fig. 11- 7.

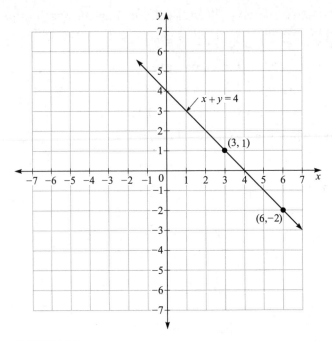

FIGURE 11-7

MATH NOTE   *If you select a different value for x, you will get a different value for y, but the point will still be on the line.*

EXAMPLE

Draw the graph of $3x - y = 7$.

SOLUTION

Find the coordinates of two points on the line.

Let $x = 1$	Let $x = 4$
$3x - y = 7$	$3x - y = 7$
$3(1) - y = 7$	$3(4) - y = 7$
$3 - y = 7$	$12 - y = 7$
$3 - 3 - y = 7 - 3$	$12 - 12 - y = 7 - 12$
$-y = 4$	$-y = -5$

$$\frac{-\overset{1}{\cancel{1}}y}{-\underset{1}{\cancel{1}}} = \frac{4}{-1} \qquad \frac{-\overset{1}{\cancel{1}}y}{-\underset{1}{\cancel{1}}} = \frac{-5}{-1}$$

$y = -4$	$y = 5$
$(1, -4)$	$(4, 5)$

Plot the two points (1, –4) and (4, 5) and draw the line through these points. See Fig. 11-8.

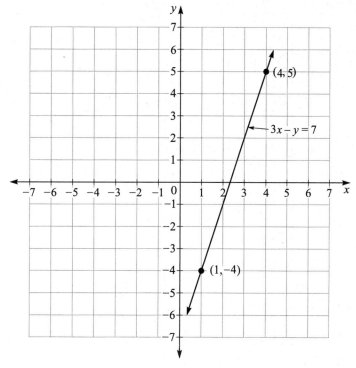

FIGURE 11-8

MATH NOTE *When graphing lines, it is best to select three points rather than two points in case an error has been made. If an error has been made, the three points will not line up.*

Practice

Draw the graph of each of the following. (Select your own points.)

1. $3x + 2y = 12$

2. $x - y = 6$

3. $-x + 5y = 16$

4. $2x - y = 8$

5. $4x + 3y = -8$

Answers

1. See Fig. 11-9.
2. See Fig. 11-10.
3. See Fig. 11-11.
4. See Fig. 11-12.
5. See Fig. 11-13.

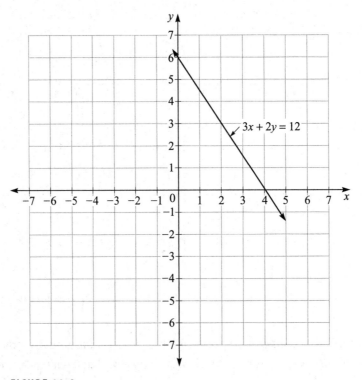

FIGURE 11-9

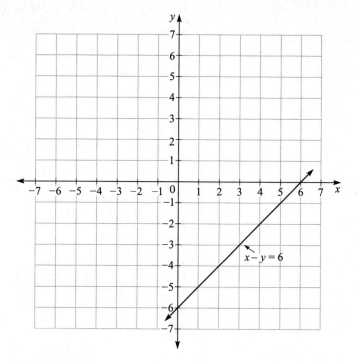

FIGURE 11-10

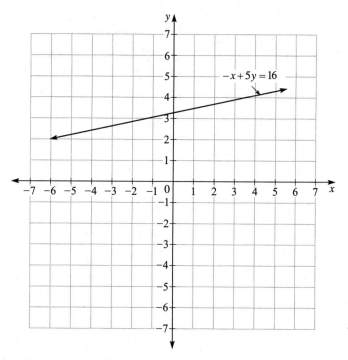

FIGURE 11-11

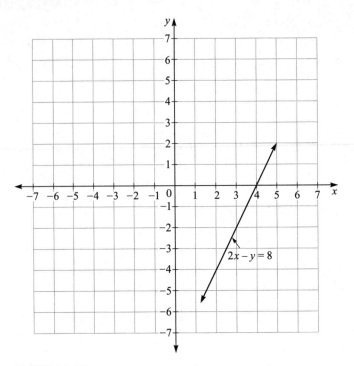

FIGURE 11-12

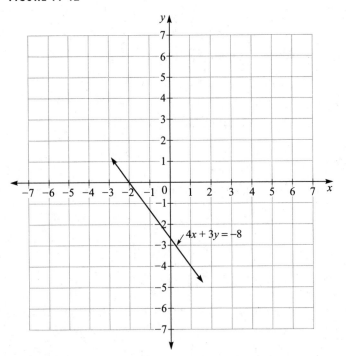

FIGURE 11-13

Horizontal and Vertical Lines

Any equation of the form x = a where a is a real number is the graph of a vertical line passing through the point (a, 0) on the x axis.

EXAMPLE

Graph the line x = –4.

✔ SOLUTION

Draw a vertical line passing through the point (–4, 0) on the x axis. See Fig. 11-14.

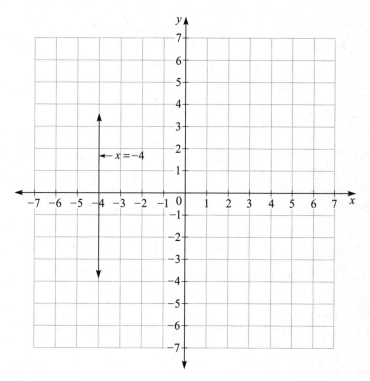

FIGURE 11-14

Any equation of the form y = b where b is any real number is the graph of a horizontal line passing through the point (0, b) on the y axis.

EXAMPLE

Graph the line y = 2.

SOLUTION

Draw a horizontal line through the point (0, 2) on the y axis. See Fig. 11-15.

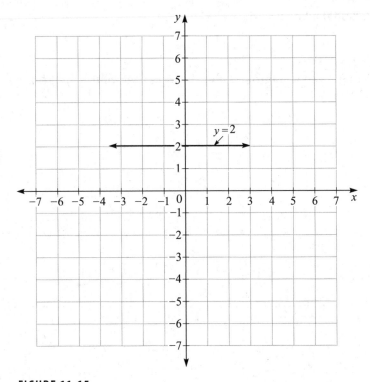

FIGURE 11-15

Practice

Draw the graph of each of the following:

1. y = –2
2. x = –5
3. y = 4
4. x = 2
5. y = 3

Answers

1. See Fig. 11-16.
2. See Fig. 11-17.
3. See Fig. 11-18.
4. See Fig. 11-19.
5. See Fig. 11-20.

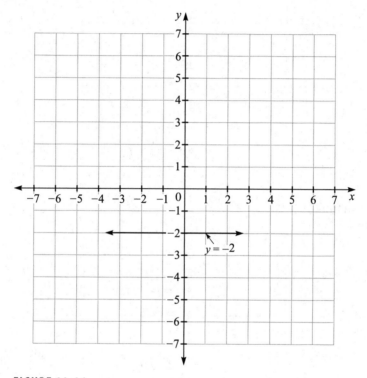

FIGURE 11-16

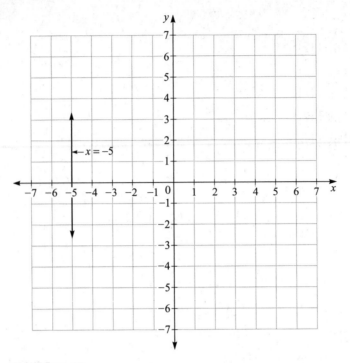

FIGURE 11-17

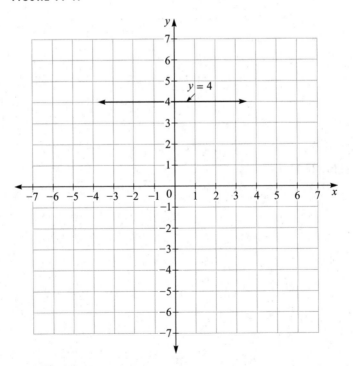

FIGURE 11-18

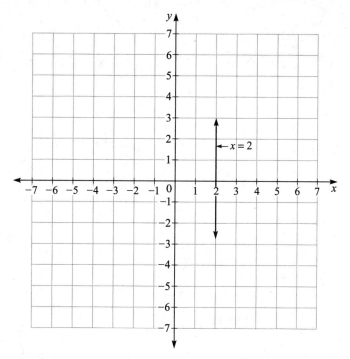

FIGURE 11-19

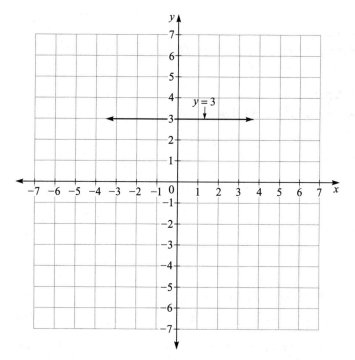

FIGURE 11-20

Intercepts

The point on the graph where a line crosses the y axis is called the **y intercept**. See Fig. 11-21.

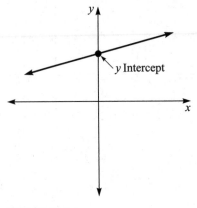

y Intercept

FIGURE 11-21

To find the y intercept, let x = 0, substitute in the equation, and then solve for y.

EXAMPLE

Find the y intercept of 4x + 3y = 12.

SOLUTION

Substitute 0 for x and solve for y.

$$4x + 3y = 12$$
$$4(0) + 3y = 12$$
$$0 + 3y = 12$$
$$\frac{\cancel{3}y}{\cancel{3}} = \frac{12}{3}$$
$$y = 4$$

Hence, the y intercept is (0, 4).

The point on the graph where a line crosses the x axis is called the **x intercept**. See Fig. 11-22.

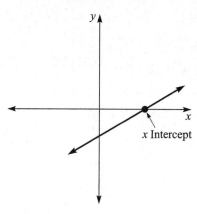

FIGURE 11-22

To find the x intercept, let y = 0, substitute in the equation, and then solve for x.

EXAMPLE

Find the x intercept for 2x + 5y = 20.

SOLUTION

Substitute 0 for y and solve for x.

$$2x + 5y = 20$$
$$2x + 5(0) = 20$$
$$2x + 0 = 20$$
$$\frac{\overset{1}{\cancel{2}}x}{\underset{1}{\cancel{2}}} = \frac{20}{2}$$
$$x = 10$$

Hence, the x intercept is (10, 0).

 Still Struggling

It is easy to draw lines on a graph using the intercepts for the two points.

Practice

1. Find the x intercept of $7x + 4y = 14$.

2. Find the y intercept of $3x - 5y = 15$.

3. Find the y intercept of $6x + 3y = 21$.

4. Find the x intercept of $5x - 6y = 20$.

5. Find the y intercept of $x - 2y = 10$.

Answers

1. $(2, 0)$
2. $(0, -3)$
3. $(0, 7)$
4. $(4, 0)$
5. $(0, -5)$

Slope

An important concept associated with lines is called the *slope* of a line. The slope of a line is associated with the "steepness" of a line. The **slope** of a line is the ratio of the vertical change to the horizontal change of a line when going from left to right. The slope is loosely defined as the rise divided by the run. See Fig. 11-23.

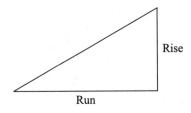

Rise

Run

FIGURE 11-23

The slope of a line can be found in several ways. A line going uphill from left to right has a slope that is positive. A line going downhill from left to right has a slope that is negative. The slope of a horizontal line is zero and the slope of a vertical line is *undefined*. See Fig. 11-24.

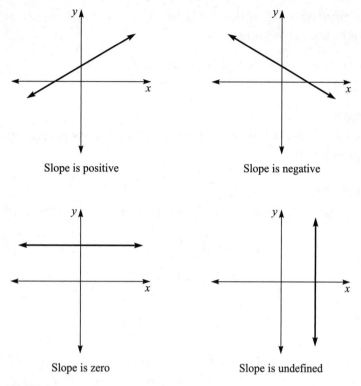

Slope is positive   Slope is negative

Slope is zero   Slope is undefined

FIGURE 11-24

On a graph the slope can be found by selecting two points, forming a right triangle, counting the number of units in the rise and run, and then dividing those two values. See Fig. 11-25.

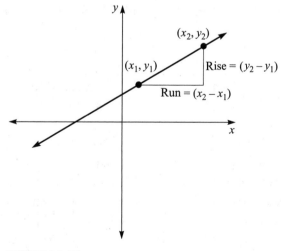

FIGURE 11-25

A better method is to find the coordinates of two points on the line, say (x_1, y_1) and (x_2, y_2), then use the formula:

$$\text{slope } (m) = \frac{y_2 - y_1}{x_2 - x_1}$$

 EXAMPLE

Find the slope of a line passing through the points whose coordinates are (4, 7) and (6, 1).

 SOLUTION

Let $x_1 = 4$, $y_1 = 7$, and $x_2 = 6$, $y_2 = 1$, and then substitute in the slope formula.

$$m = \frac{y_2 - y_1}{x_2 - x_1}$$

$$= \frac{1 - 7}{6 - 4}$$

$$= \frac{-6}{2} = -\frac{3}{1} \text{ or } -3$$

Hence, the slope of the line is –3.

The slope of a line whose equation is known can be found by selecting any two points on the line and then using the slope formula.

 EXAMPLE

Find the slope of a line whose equation is $3x + 5y = 10$.

 SOLUTION

Select two points on the line.

Let x = 5, then	Let x = –10, then
$3x + 5y = 10$	$3x + 5y = 10$
$3(5) + 5y = 10$	$3(-10) + 5y = 10$
$15 + 5y = 10$	$-30 + 5y = 10$
$15 - 15 + 5y = 10 - 15$	$-30 + 30 + 5y = 10 + 30$
$5y = -5$	$5y = 40$
$\dfrac{\cancel{5}y}{\cancel{5}} = \dfrac{-5}{5}$	$\dfrac{\cancel{5}y}{\cancel{5}} = \dfrac{40}{5}$
$y = -1$	$y = 8$
(5, –1)	(–10, 8)

Let $x_1 = 5$, $y_1 = -1$, and $x_2 = -10$, $y_2 = 8$.

Now substitute in the slope formula:

$$m = \frac{y_2 - y_1}{x_2 - x_1}$$

$$m = \frac{8 - (-1)}{-10 - 5}$$

$$= \frac{9}{-15}$$

$$= -\frac{3}{5}$$

The slope of the line $3x + 5y = 10$ is $-\frac{3}{5}$.

MATH NOTE *If you select different numbers for x, you will get different numbers for y; however, the slope will always be the same.*

Another way to find the slope of a line is to solve the equation for y in terms of x. The coefficient of x will be the slope.

 EXAMPLE

Find the slope of a line whose equation is $3x + 5y = 10$.

 SOLUTION

Solve the equation for y as shown:

$$3x + 5y = 10$$

$$3x - 3x + 5y = -3x + 10$$

$$5y = -3x + 10$$

$$\frac{\cancel{5}y}{\cancel{5}} = \frac{-3x}{5} + \frac{10}{5}$$

$$y = -\frac{3}{5}x + 2$$

Hence, the slope is $-\frac{3}{5}$ which is the result found in the previous example.

Still Struggling

An equation in the form of y in terms of x is called the slope-intercept form and in general is written as $y = mx + b$, where m is the slope and b is the y intercept.

Practice

1. Find the slope of the line containing the two points $(-8, -4)$ and $(3, 1)$.
2. Find the slope of the line containing the two points $(2, -4)$ and $(3, 6)$.
3. Find the slope of the line whose equation is $7x - 2y = 9$.
4. Find the slope of the line whose equation is $-3x + 4y = -11$.
5. Find the slope of the line whose equation is $x + 4y = 16$.

Answers

1. $\dfrac{5}{11}$

2. 10

3. $\dfrac{7}{2}$

4. $\dfrac{3}{4}$

5. $-\dfrac{1}{4}$

Solving a System of Linear Equations

Two linear equations of the form $ax + by = c$, where a, b, and c are real numbers, is called a **system** of linear equations. When the two lines intersect, the coordinates of the point of intersection are called the **solution** of the system. The system then is said to be **independent** and **consistent**. When the lines are parallel, there is no point of intersection; hence, there is no solution for the system.

In this case the system is said to be **inconsistent.** When the two lines coincide, every point on the line then is a solution. The system is said to be **dependent.** See Fig. 11-26.

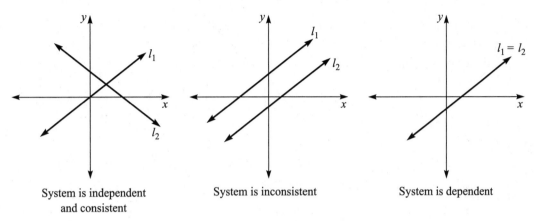

System is independent   System is inconsistent   System is dependent
and consistent

FIGURE 11-26

To find a solution to a system of linear equations, plot the graphs for the lines and find the point of intersection.

 EXAMPLE

Find the solution of the system.

$$x + y = 6$$

$$x - y = -2$$

 SOLUTION

Find two points on each line. For $x + y = 6$, use (5, 1) and (3, 3). For $x - y = -2$, use (−1, 1) and (3, 5). Graph both lines and then find the point of intersection. See Fig. 11-27.

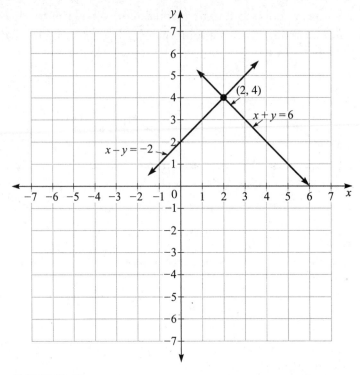

FIGURE 11-27

The point of intersection is (2, 4).

MATH NOTE   *To check, substitute the values of x and y for the solution in both equations and see if they are closed true equations.*

 EXAMPLE
Find the solution of the system.

$$2x + y = 3$$
$$x - 2y = 9$$

✔ **SOLUTION**
Find two points on each line. For $2x + y = 3$, use (4, –5) and (1, 1). For $x - 2y = 9$, use (7, –1) and (5, –2). Graph both lines and then find the point of intersection. See Fig. 11-28.

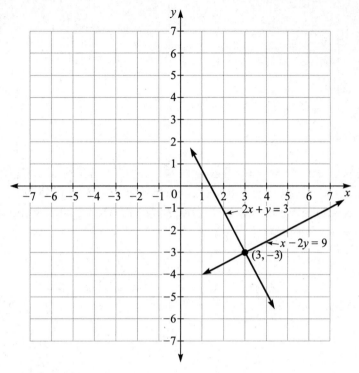

FIGURE 11-28

The point of intersection is (3, –3).

Practice

Find the solution for each system of equations:

1. x + y = 7
   x – y = 3

2. 3x + y = 8
   x + 2y = 11

3. 7x + y = 16
   2x – 3y = –2

4. x – 6y = 9
   4x – y = 13

5. x + 4y = 10
   6x – y = –15

Answers

1. (5, 2)
2. (1, 5)
3. (2, 2)
4. (3, –1)
5. (–2, 3)

In this chapter, you have learned how to graph points and lines. Each line, except a vertical line, has a slope. Lines also have intercepts. Finally, a system of linear equations can be solved by graphing the two equations and finding the point of intersection.

QUIZ

Use Fig. 11-29 to answer questions 1–5:

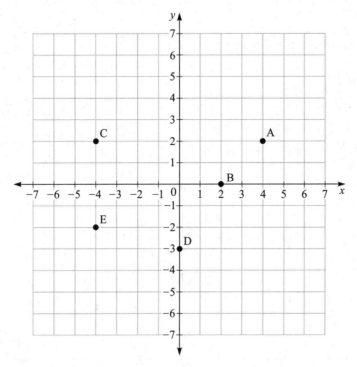

FIGURE 11-29

1. The point whose coordinates are (0, −3) is:
   A. point A
   B. point B
   C. point D
   D. point E

2. The point whose coordinates are (4, 2) is:
   A. point A
   B. point B
   C. point C
   D. point E

3. The point whose coordinates are (−4, −2) is:
   A. point B
   B. point C
   C. point D
   D. point E

4. **The point whose coordinates are (2, 0) is:**
   A. point A
   B. point B
   C. point C
   D. point D

5. **The point whose coordinates are (–4, 2) is:**
   A. point A
   B. point B
   C. point C
   D. point E

Use Fig. 11-30 to answer questions 6–9.

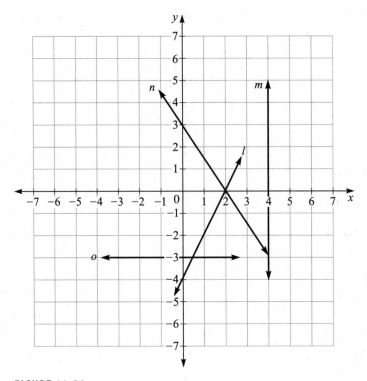

FIGURE 11-30

6. **The line whose equation is 3x + 2y = 6 is:**
   A. line l
   B. line m
   C. line n
   D. line o

7. **The line whose equation is $x = 4$ is:**
   A. line l
   B. line m
   C. line n
   D. line o

8. **The line whose equation is $2x - y = 4$ is:**
   A. line l
   B. line m
   C. line n
   D. line o

9. **The line whose equation is $y = -3$ is:**
   A. line l
   B. line m
   C. line n
   D. line o

10. **Which is a solution to $4x - y = 8$?**
    A. $(2, 2)$
    B. $(-2, 0)$
    C. $(3, -4)$
    D. $(0, -8)$

11. **Which is a solution to $-3x + 2y = 6$?**
    A. $(2, 0)$
    B. $(-4, 9)$
    C. $(4, 9)$
    D. $(0, -3)$

12. **Find y when $x = -3$ for $4x + 5y = -12$.**
    A. 5
    B. 0
    C. -2
    D. -3

13. **Find x when $y = -3$ for $x + 3y = -10$.**
    A. $x = 1$
    B. $x = 2$
    C. $x = -1$
    D. $x = -2$

14. Find the x intercept for −6x + 5y = 30.

   A. (−5, 0)
   B. (0, 6)
   C. (0, −6)
   D. (5, 0)

15. Find the y intercept for 3x − 7y = 21.

   A. (7, 0)
   B. (−3, 0)
   C. (0, −3)
   D. (−7, 0)

16. Find the slope of a line containing the two points whose coordinates are (5, 7) and (−2, 2).

   A. $\dfrac{7}{5}$

   B. $-\dfrac{5}{7}$

   C. $-\dfrac{7}{5}$

   D. $\dfrac{5}{7}$

17. Find the slope of the line −5x + 2y = 10.

   A. $\dfrac{2}{5}$

   B. $\dfrac{5}{2}$

   C. $-\dfrac{2}{5}$

   D. −2

18. The slope of a horizontal line is

   A. 0
   B. 1
   C. undefined
   D. −1

19. When two lines are parallel, the system is said to be:

   A. inconsistent
   B. dependent
   C. undefined
   D. consistent

20. **Which point is a solution for the system?**

$$3x - y = 10$$
$$2x + y = -5$$

A. $(4, 2)$
B. $(1, 3)$
C. $(1, -7)$
D. $(-5, 0)$

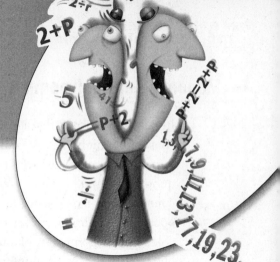

chapter 12

Operations with Monomials and Polynomials

This chapter explains the basic operations of addition, subtraction, multiplication, and division of algebraic expressions. If you learn these concepts now, it will make the next course easier since they are the basic concepts that are used in algebra.

CHAPTER OBJECTIVES

In this chapter, you will learn how to

- Identify monomials and polynomials
- Add and subtract polynomials
- Multiply monomials
- Raise a monomial to a power
- Multiply binomials using the FOIL method
- Square a binomial
- Multiply polynomials
- Divide monomials
- Divide a polynomial by a monomial

Monomials and Polynomials

Recall from Chap. 7 that an **algebraic expression** consists of variables (letters), constants (numbers), operation signs, and grouping symbols. Also recall that a **term** of an algebraic expression consists of a number, variable, or a product or quotient of numbers and variables. The terms of an algebraic expression are connected by + or − signs. An algebraic expression consisting only of the four operations (addition, subtraction, multiplication, division), which has no variable in the denominator of a term, is called a **polynomial**. If the expression has one term, it is called a **monomial**. If the expression has two terms, it is called a **binomial**. If the expression has three terms, it is called a **trinomial**.

Monomials:	$3x^2$	$-5y$	$2x^3y^2$
Binomials:	$x + 2y$	$3x^2 + x$	$-2y + z$
Trinomials:	$3x^2 + 4x - 1$		$2x - 3y + 5$

Addition of Polynomials

Recall from Chap. 7 that only like terms can be added or subtracted. For example, $5x + 6x = 11x$. Unlike terms cannot be added or subtracted.

To add two or more polynomials, add like terms.

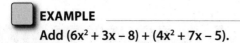 **EXAMPLE**

Add $(6x^2 + 3x - 8) + (4x^2 + 7x - 5)$.

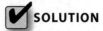 **SOLUTION**

$$(6x^2 + 3x - 8) + (4x^2 + 7x - 5) = 6x^2 + 3x - 8 + 4x^2 + 7x - 5$$
$$= (6x^2 + 4x^2) + (3x + 7x) + (-8 - 5)$$
$$= 10x^2 + 10x - 13$$

MATH NOTE   *Recall that when the numerical coefficient is one, it is not written. That is, $xy = 1xy$. Likewise $-xy = -1xy$.*

 **EXAMPLE**

Add $(8x + 2y) + (3x - 7) + (2y - 6)$.

 SOLUTION

$$(8x + 2y) + (3x - 7) + (2y - 6) = 8x + 2y + 3x - 7 + 2y - 6$$
$$= (8x + 3x) + (2y + 2y) + (-7 - 6)$$
$$= 11x + 4y - 13$$

Still Struggling

If the sum of two like terms is zero, do not write the term in the sum.

Practice

Add:

1. $(9x^2 + 2x + 6) + (5x^2 - 6x - 8)$
2. $(4x + 7y - 3) + (4x - 3y + 10)$
3. $(4x + 7) + (3x - 8) + (2x - 6) + (5x - 10)$
4. $(3a^2b^2 - 6ab - 7) + (a^2b^2 + 4ab + 11)$
5. $(9m + 2n - 6) + (3n - 1) + (6m + 15)$

Answers

1. $14x^2 - 4x - 2$
2. $8x + 4y + 7$
3. $14x - 17$
4. $4a^2b^2 - 2ab + 4$
5. $15m + 5n + 8$

Subtraction of Polynomials

In Chap. 2 you learned that whenever you subtract integers in algebra, you add the opposite. For example, $-10 - (-6) = -10 + 6 = -4$. To find the opposite of an integer (except 0), we change its sign. To find the opposite of a monomial, change the sign of the numerical coefficient. For example, the opposite of $-7xy$ is $7xy$. The opposite of $8x^2$ is $-8x^2$.

To find the opposite of a polynomial, change the sign of every term of the polynomial. For example, the opposite of $6x^2 - 3x + 2$ is $-(6x^2 - 3x + 2)$ or $-6x^2 + 3x - 2$.

To subtract two polynomials, add the opposite of the polynomial being subtracted.

EXAMPLE

Subtract $(12x^2 - 5x + 6) - (8x^2 - 3x - 2)$.

SOLUTION

$$(12x^2 - 5x + 6) - (8x^2 - 3x - 2) = 12x^2 - 5x + 6 - 8x^2 + 3x + 2$$
$$= (12x^2 - 8x^2) + (-5x + 3x) + (6 + 2)$$
$$= 4x^2 - 2x + 8$$

EXAMPLE

Subtract $(12a - 10b + c) - (9a + 3b - c)$.

SOLUTION

$$(12a - 10b + c) - (9a + 3b - c) = 12a - 10b + c - 9a - 3b + c$$
$$= (12a - 9a) + (-10b - 3b) + (c + c)$$
$$= 3a - 13b + 2c$$

Practice
Subtract:

1. $(5y - 7) - (2y + 6)$
2. $(3y^2 + 7y - 3) - (4y^2 - 6y + 10)$

3. $(6a + 3b - 10) - (-5a + 7b - 9)$

4. $(5r + 7s) - (9s + 3r)$

5. $(x^2 - 3x - 11) - (7 + 2x - x^2)$

Answers

1. $3y - 13$

2. $-y^2 + 13y - 13$

3. $11a - 4b - 1$

4. $2r - 2s$

5. $2x^2 - 5x - 18$

Multiplication of Monomials

When two exponential expressions with the same base are multiplied, the exponents are added to get the product. For example,

$$x^3 \cdot x^4 = x \cdot x \cdot x \cdot x \cdot x \cdot x \cdot x = x^{3+4} = x^7$$

In general, $x^m \cdot x^n = x^{m+n}$.

 EXAMPLE

Multiply $y^4 \cdot y^7$.

 SOLUTION

$$y^4 \cdot y^7 = y^{4+7} = y^{11}$$

To multiply two monomials, multiply the numerical coefficients (numbers) and add the exponents of the same variables (letters).

 EXAMPLE

Multiply $4x^4y^3 \cdot 2xy^6$.

 **SOLUTION**

$$4x^4y^3 \cdot 2xy^6 = 4 \cdot 2 \cdot x^4 \cdot x \cdot y^3 \cdot y^6$$
$$= 8x^5y^9$$

EXAMPLE

Multiply $-5m^3n^5 \cdot 6m \cdot n^4$.

SOLUTION

$$-5m^3n^5 \cdot 6m \cdot n^4 = -5 \cdot 6 \cdot m^3 \cdot m \cdot n^5 \cdot n^4$$

$$= -30m^4n^9$$

Still Struggling

Recall that $x = x^1$.

EXAMPLE

Multiply $-x^2y^3 \cdot 2xy^3$.

SOLUTION

$$-x^2y^3 \cdot 2xy^3 = -1x^2y^3 \cdot 2xy^3$$

$$= -1 \cdot 2 \cdot x^2 \cdot x \cdot y^3 \cdot y^3$$

$$= -2x^3y^6$$

Practice

Multiply:

1. $5pq \cdot 7pq$
2. $-3x^4 \cdot 2x^2y^4$
3. $3a^3b \cdot 5a^4b^2 \cdot 6ab$
4. $-11xy^5 \cdot 2xy^2 \cdot 3xy$
5. $2xz \cdot 5x^2y \cdot (-4y^3z^2)$

Answers

1. $35p^2q^2$
2. $-6x^6y^4$
3. $90a^8b^4$
4. $-66x^3y^8$
5. $-40x^3y^4z^3$

Raising a Monomial to a Power

When a variable is raised to a power, the exponent of the variable is multiplied by the power. For example,

$$(x^2)^3 = x^2 \cdot x^2 \cdot x^2 = x^{2+2+2} = x^6$$

or

$$(x^2)^3 = x^{2 \cdot 3} = x^6$$

In general $(x^m)^n = x^{m \cdot n}$. To raise a monomial to a power, raise the numerical coefficient to the power and multiply the exponents of the variables by the power.

 EXAMPLE

Find $(6x^4y^2)^3$.

 **SOLUTION**

$$(6x^4y^2)^3 = 6^3 \cdot x^{4 \cdot 3} \, y^{2 \cdot 3}$$
$$= 216x^{12}y^6$$

EXAMPLE

Find $(-4a^5b^2)^3$.

SOLUTION

$$(-4a^5b^2)^3 = (-4)^3 a^{5 \cdot 3} b^{2 \cdot 3}$$
$$= -64a^{15}b^6$$

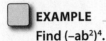
EXAMPLE
Find $(-ab^2)^4$.

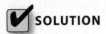
SOLUTION

$$(-ab^2)^4 = (-1)^4 \, a^{1 \cdot 4} b^{2 \cdot 4}$$
$$= 1a^4 b^8$$
$$= a^4 b^8$$

Practice

Find each of the following:

1. $(y^3)^6$
2. $(5b)^3$
3. $(3m^3 n^5)^3$
4. $(-2a^4 b^2)^2$
5. $(9x^4 y^2)^3$

Answers

1. y^{18}
2. $125b^3$
3. $27m^9 n^{15}$
4. $4a^8 b^4$
5. $729x^{12} y^6$

Multiplication of a Polynomial by a Monomial

When a polynomial is multiplied by a monomial, the distributive property is used. Recall from Chap. 7 that the distributive property states that $a(b + c) = a \cdot b + a \cdot c$.

To multiply a polynomial by a monomial, multiply each term in the polynomial by the monomial.

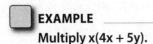
EXAMPLE
Multiply $x(4x + 5y)$.

SOLUTION

$$x(4x + 5y) = x \cdot 4x + x \cdot 5y$$

$$= 4x^2 + 5xy$$

EXAMPLE

Multiply $2x^3y^2(6x + 8y - 4)$.

 **SOLUTION**

$$2x^3y^2(6x + 8y - 4) = 2x^3y^2 \cdot 6x + 2x^3y^2 \cdot 8y - 2x^3y^2 \cdot 4$$

$$= 12x^4y^2 + 16x^3y^3 - 8x^3y^2$$

Practice

Multiply:

1. $5(6a - 2b + 3)$
2. $3y(6y^2 - 6y + 18)$
3. $-2cd^3(3cd + 4c^3 - 5)$
4. $-x^2y^4(2x + 3xy - 8y^5)$
5. $3pq(2p - 6q + 7r^2)$

Answers

1. $30a - 10b + 15$
2. $18y^3 - 18y^2 + 54y$
3. $-6c^2d^4 - 8c^4d^3 + 10cd^3$
4. $-2x^3y^4 - 3x^3y^5 + 8x^2y^9$
5. $6p^2q - 18pq^2 + 21pqr^2$

Multiplication of Two Binomials

In algebra, there are some special products that you will need to know. The first one is to be able to find the product of two binomials. In order to do this, you can use the distributive law twice.

EXAMPLE

Multiply $(x + 6)(x + 3)$.

SOLUTION

Distribute $(x + 6)$ over $(x + 3)$ as shown:

$$(x + 6)(x + 3) = (x + 6)x + (x + 6)3.$$

Next distribute again as shown:

$$= x \cdot x + 6x + 3 \cdot x + 6 \cdot 3$$
$$= x^2 + 6x + 3x + 18$$

Then combine like terms:

$$= x^2 + 9x + 18$$

Hence, $(x + 6)(x + 3) = x^2 + 9x + 18$.

EXAMPLE

Multiply $(2x - 5)(x + 6)$.

SOLUTION

$$(2x - 5)(x + 6) = (2x - 5)x + (2x - 5)6$$
$$= 2x \cdot x - 5 \cdot x + 2x \cdot 6 - 5 \cdot 6$$
$$= 2x^2 - 5x + 12x - 30$$
$$= 2x^2 + 7x - 30$$

A shortcut method for multiplying two binomials is called the FOIL method. FOIL stands for:

F = First   O = Outer   I = Inner   L = Last

Using the FOIL method means finding the product of the first terms in each factor, the product of the outer terms in each factor, the product of the inner terms in each factor, the product of the last terms in each factor, and then finding the sum of these four products.

EXAMPLE

Multiply $(x - 8)(x + 3)$ using the FOIL method.

 **SOLUTION**

First $x \cdot x = x^2$
Outer $x \cdot 3 = 3x$
Inner $-8 \cdot x = -8x$
Last $-8 \cdot 3 = -24$

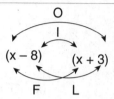

Add the like terms: $3x + (-8x) = -5x$.
Hence, $(x - 8)(x + 3) = x^2 - 5x - 24$

Practice
Multiply:

1. $(x + 7)(x + 9)$
2. $(x - 8)(x - 7)$
3. $(2x + 3)(x - 12)$
4. $(x - 5)(x + 5)$
5. $(x + 7)(x + 7)$

Answers

1. $x^2 + 16x + 63$
2. $x^2 - 15x + 56$
3. $2x^2 - 21x - 36$
4. $x^2 - 25$
5. $x^2 + 14x + 49$

Squaring a Binomial

Another special product results from squaring a binomial. This can be done by using the two methods shown previously; however, a shortcut rule can be used. It is

$$(a + b)^2 = a^2 + 2ab + b^2$$

$$(a - b)^2 = a^2 - 2ab + b^2$$

In words, whenever you square a binomial, square the first term and then multiply the product of the first and second terms by 2 and square the last term.

EXAMPLE
Find $(x + 7)^2$.

SOLUTION

$$(x + 7)^2 = x^2 + 2 \cdot 7 \cdot x + 7^2$$
$$= x^2 + 14x + 49$$

EXAMPLE
Find $(3x - 5)^2$.

SOLUTION

$$(3x - 5)^2 = (3x)^2 + 2 \cdot 3x(-5) + (-5)^2$$
$$= 9x^2 - 30x + 25$$

EXAMPLE
Find $(5y - 1)^2$.

SOLUTION

$$(5y - 1)^2 = (5y)^2 + 2 \cdot 5y \cdot (-1) + (-1)^2$$
$$= 25y^2 - 10y + 1$$

Practice

1. $(y + 2)^2$
2. $(y - 8)^2$
3. $(3x + 4)^2$
4. $(6x + 7)^2$
5. $(2x - 9)^2$

Answers

1. $y^2 + 4y + 4$

2. $y^2 - 16y + 64$

3. $9x^2 + 24x + 16$

4. $36x^2 + 84x + 49$

5. $4x^2 - 36x + 81$

Multiplication of Two Polynomials

Two polynomials can be multiplied by using the distributive property as many times as needed.

 EXAMPLE

Multiply $(x + 6)(x^2 - 8x + 9)$.

 **SOLUTION**

$$(x + 6)(x^2 - 8x + 9) = (x + 6)x^2 - (x + 6)8x + (x + 6)9$$
$$= x \cdot x^2 + 6 \cdot x^2 - x \cdot 8x - 6 \cdot 8x + x \cdot 9 + 6 \cdot 9$$
$$= x^3 + 6x^2 - 8x^2 - 48x + 9x + 54$$
$$= x^3 - 2x^2 - 39x + 54$$

Multiplication of polynomials can be performed vertically. The process is similar to multiplying whole numbers.

 EXAMPLE

Multiply $(x + 6)(x^2 - 8x + 9)$ vertically.

 **SOLUTION**

$$
\begin{array}{r}
x^2 - 8x + 9 \\
x + 6 \\
\hline
6x^2 - 48x + 54 \\
x^3 - 8x^2 + 9x \\
\hline
x^3 - 2x^2 - 39x + 54
\end{array}
$$

Multiply the top by 6
Multiply the top by x
Add like terms

EXAMPLE

Multiply $(x^2 - 5x - 9)(3x - 8)$ vertically.

SOLUTION

$$
\begin{array}{r}
x^2 - 5x - 9 \\
3x - 8 \\
\hline
-8x^2 + 40x + 72 \\
3x^3 - 15x^2 - 27x \\
\hline
3x^3 - 23x^2 + 13x + 72
\end{array}
$$

MATH NOTE *Be sure to place like terms under each other when multiplying vertically.*

Practice

Multiply:

1. $(x^2 + 4x + 3)(x + 5)$
2. $(4x^2 + x - 6)(x - 1)$
3. $(3x^2 + 4x - 6)(4x - 7)$
4. $(x^3 + 3x^2 + 4x + 1)(x + 4)$
5. $(x^2 + 3x - 6)(x^2 + x - 5)$

Answers

1. $x^3 + 9x^2 + 23x + 15$
2. $4x^3 - 3x^2 - 7x + 6$
3. $12x^3 - 5x^2 - 52x + 42$
4. $x^4 + 7x^3 + 16x^2 + 17x + 4$
5. $x^4 + 4x^3 - 8x^2 - 21x + 30$

Division of Monomials

When one variable is divided by another variable with the same base, the exponent of the variable in the denominator is subtracted from the exponent of the variable in the numerator. For example,

$$
\frac{x^6}{x^2} = \frac{x \cdot x \cdot x \cdot x \cdot x \cdot x}{x \cdot x} = x^{6-2} = x^4
$$

In general, $\dfrac{x^m}{x^n} = x^{m-n}$.

Since division by zero is undefined, all variable in the dnominator of the fraction of the examples and exercises cannot be zero.

EXAMPLE

Divide x^8 by x^5.

SOLUTION

$$\frac{x^8}{x^5} = x^{8-5} = x^3$$

To divide a monomial by a monomial, divide the numerical coefficients and then subtract the exponents of the same variables.

EXAMPLE

Divide $20x^6$ by $-4x^2$.

SOLUTION

$$\frac{20x^6}{-4x^2} = -5x^{6-2} = -5x^4$$

EXAMPLE

Divide $18x^6y^4$ by $9xy^2$.

SOLUTION

$$\frac{18x^6y^4}{9xy^2} = 2x^{6-1}y^{4-2} = 2x^5y^2$$

MATH NOTE $x^0 = 1\ (x \neq 0)$

EXAMPLE

Divide $24x^5y^3z^4$ by $8xy^3z$.

SOLUTION

$$\frac{24x^5y^3z^4}{8xy^3z} = 3x^{5-1}y^{3-3}z^{4-1} = 3x^4z^3$$

Practice
Divide:

1. $-6x^4y^7 \div 2xy^3$
2. $42m^2n^4 \div (-6m^2n^3)$
3. $24c \div 8$
4. $16c^4d^3 \div 2cd^2$
5. $51a^5b^3 \div 3a^4b^2$

Answers

1. $-3x^3y^4$
2. $-7n$
3. $3c$
4. $8c^3d$
5. $17ab$

Division of a Polynomial by a Monomial

To divide a polynomial by a monomial, divide each term in the polynomial by the monomial.

 EXAMPLE
Divide $12x^2 + 6x$ by $3x$.

 SOLUTION

$$\frac{12x^2 + 6x}{3x} = \frac{12x^2}{3x} + \frac{6x}{3x} = 4x + 2$$

 EXAMPLE
Divide $15x^4y^3 + 10x^2y^5 \div (-5xy)$.

 SOLUTION

$$\frac{15x^4y^3 + 10x^2y^5}{-5xy} = \frac{15x^4y^3}{-5xy} + \frac{10x^2y^5}{-5xy} = -3x^3y^2 - 2xy^4$$

EXAMPLE

Divide $27x^5 + 24x^3 - 18x$ by 3.

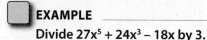 **SOLUTION**

$$\frac{27x^5 + 24x^3 - 18x}{3} = \frac{27x^5}{3} + \frac{24x^3}{3} + \frac{-18x}{3} = 9x^5 + 8x^3 - 6x$$

Practice
Divide:

1. $(42x^2 - 36x) \div 6x$
2. $(15x^2 - 10x - 5) \div 5$
3. $(24c^3d^4 + 18c^2d^2) \div (-6c)$
4. $(5a^3b + 10a^2b + 5ab) \div ab$
5. $(12x^4y^6 - 8x^6y^2) \div (-4xy)$

Answers

1. $7x - 6$
2. $3x^2 - 2x - 1$
3. $-4c^2d^4 - 3cd^2$
4. $5a^2 + 10a + 5$
5. $-3x^3y^5 + 2x^5y$

In this chapter, you have learned how to add, subtract, multiply, and divide monomial and polynomial expressions. These concepts are used extensively in the next course in algebra. If you have a good understanding of this material along with the other material presented in Chap. 2, Chap. 7, and Chap. 11, you will have a good foundation for the next algebra course you will be taking.

QUIZ

1. **Add $(4x^2 + 2x - 1) + (9x^2 - 8x - 3)$.**
   A. $13x^2 - 4x + 2$
   B. $6x^2 - 4x + 2$
   C. $5x^2 - 5x + 2$
   D. $13x^2 - 6x - 4$

2. **Add $(12x + 6y) + (2x - 7y)$.**
   A. $10x + 10y$
   B. $14x - y$
   C. $10x - 6y$
   D. $14x + y$

3. **Subtract $(3x - 7) - (-5x + 6)$.**
   A. $8x - 13$
   B. $2x + 1$
   C. $8x - 1$
   D. $2x - 13$

4. **Subtract $(5a - 7b + c) - (4a - 6b + 3c)$.**
   A. $-12a - 10b + 6c$
   B. $9a - 13b + 4c$
   C. $a - b - 2c$
   D. $-a - 7b - 5c$

5. **Multiply $5x^3 \cdot 2x^2$.**
   A. $7x^4$
   B. $10x^6$
   C. $10x^5$
   D. $7x^5$

6. **Multiply $5x^2 \cdot 7xy \cdot 2xy^2$.**
   A. $14x^2y^2$
   B. $70x^4y^3$
   C. $35x^2y^2$
   D. $19x^4y^2$

7. **Find $(4x^2)^4$.**
   A. $256x^8$
   B. $64x^6$
   C. $12x^6$
   D. $16x^8$

8. **Find $(-3x^3y^2)^4$.**
   A. $-81x^{12}y^8$
   B. $12x^7y^6$
   C. $-12x^7y^6$
   D. $81x^{12}y^8$

9. **Multiply $y(3y^3 - y + 8)$.**
   A. $3y^4 - y^2 + 8y$
   B. $3y^3 + y^2 + 8y$
   C. $y^4 + y^2 + 8$
   D. $3y^3 - y + 8y$

10. **Multiply $-7a(2a^2 + 7ab - 9b)$.**
    A. $-14a^2 - 49ab + 63ab$
    B. $9a^3 + 49a^2b - 9ab$
    C. $-14a^3 - 49a^2b + 63ab$
    D. $-9a^3 - 14a + 16ab$

11. **Multiply $(x + 8)(2x - 5)$.**
    A. $2x^2 + 11x + 40$
    B. $2x^2 + 11x - 40$
    C. $3x^2 + 5x - 13$
    D. $3x^2 + 21x - 40$

12. **Multiply $(3x - 1)(4x + 7)$.**
    A. $12x^2 + 17x - 7$
    B. $7x^2 + 17x + 7$
    C. $12x^2 - 7$
    D. $7x^2 + 25x - 7$

13. **Find $(x + 9)^2$.**
    A. $x^2 + 81$
    B. $x^2 + 18$
    C. $x^2 + 9x + 81$
    D. $x^2 + 18x + 81$

14. **Find $(2y - 7)^2$.**
    A. $4y^2 + 49$
    B. $4y^2 + 28y + 49$
    C. $4y^2 - 28x + 49$
    D. $4y^2 - 49$

15. **Multiply $(6a^2 + 2a - 5)(3a + 1)$.**
    A. $18a^3 - 13a - 5$
    B. $18a^3 + 12a^2 - 13a - 5$
    C. $18a^3 - 12a^2 + 13a + 5$
    D. $18a^3 + 12a^2 - 17a + 1$

16. **Multiply $(y - 3)(2y^2 + y - 1)$.**
    A. $2y^3 - 7y^2 - 2y + 3$
    B. $2y^3 + 7y^2 + 2y - 3$
    C. $2y^3 - 5y^2 + 4y - 3$
    D. $2y^3 - 5y^2 - 4y + 3$

17. **Divide $20x^4 \div 4x^2$.**
    A. 5
    B. $5x^2$
    C. $5x$
    D. $5x^3$

18. **Divide $-42a^5b^3c^4 \div 2a^3b^2c$.**
    A. $-21a^2bc^3$
    B. $21a^2b^2c$
    C. $-21abc^2$
    D. $21a^2 b^2c^3$

19. **Divide $27y^3 + 30y^2 + 6$ by 3.**
    A. $9y^3 + 30y^2 - 2$
    B. $9y^3 - 10y^2 - 2$
    C. $9y^3 + 10y^2 + 2$
    D. $9y^3 - 10y^2 - 6$

20. **Divide $25a^4b^3 - 15a^2b^3 - 10ab$ by $-5ab$.**
    A. $-5a^4b^3 + 3a^2b - 2$
    B. $-5a^3b^2 + 3ab^2 + 2$
    C. $5a^3b^2 - 3ab^2 - 2$
    D. $5a^3b^2 - 3ab^2 - 1$

Final Exam

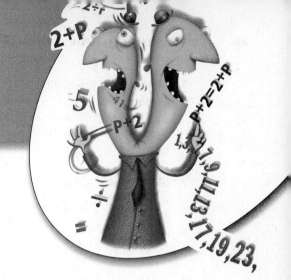

1. **Name 8,263,005.**
   A. eight million, two hundred sixty-three thousand, five
   B. eight thousand, five million
   C. eight billion, two hundred sixty-three thousand, five
   D. eight billion two hundred sixty-three million, five

2. **Round 24,687 to the nearest hundred.**
   A. 24,787
   B. 24,600
   C. 25,000
   D. 24,700

3. **Add 2,463 + 711 + 34,208.**
   A. 36,355
   B. 37,382
   C. 36,459
   D. 37,206

4. **Subtract 762,001 − 9,263.**
   A. 752,738
   B. 751,238
   C. 752,783
   D. 751,327

5. **Multiply 327 × 4,182.**
   A. 1,349,856
   B. 1,376,541
   C. 1,367,514
   D. 1,555,705

6. **Divide 374,088 ÷ 572.**
   A. 645
   B. 654
   C. 546
   D. 564

7. **Divide 61,421 ÷ 72.**
   A. 853 R 5
   B. 835 R 6
   C. 835 R 5
   D. 853 R 8

8. **Find the total number of calculators in an order if there are 15 boxes with 9 calculators in each box.**
   A. 24
   B. 150
   C. 135
   D. 108

9. **Find |6|.**
   A. −6
   B. 6
   C. 0
   D. |−6|

10. **Find the opposite of −10.**
    A. |−5|
    B. 0
    C. −10
    D. 10

11. **Add −8 + 5.**
    A. −3
    B. +3
    C. 13
    D. −13

12. **Subtract −14 − (−6).**
    A. 8
    B. −20
    C. −8
    D. +20

13. **Multiply (8)(−3)(−4).**
    A. −20
    B. 96

C. –96
D. 20

14. **Divide –32 ÷ (–8).**
    A. –4
    B. –40
    C. 40
    D. 4

15. **Simplify 16 – 2 × 7 + 3².**
    A. 11
    B. 107
    C. 140
    D. –49

16. **Find (–7)³.**
    A. 343
    B. –21
    C. –343
    D. 49

17. **Reduce $\dfrac{133}{152}$ to lowest terms.**
    A. $\dfrac{5}{6}$
    B. $\dfrac{7}{8}$
    C. $\dfrac{3}{4}$
    D. $\dfrac{19}{20}$

18. **Change $\dfrac{4}{7}$ to an equivalent fraction in higher terms.**
    A. $\dfrac{11}{21}$
    B. $\dfrac{27}{49}$
    C. $\dfrac{15}{28}$
    D. $\dfrac{24}{42}$

19. Change $\dfrac{15}{4}$ to a mixed number.

A. $3\dfrac{3}{4}$

B. $1\dfrac{1}{4}$

C. $3\dfrac{4}{5}$

D. $2\dfrac{3}{4}$

20. Change $8\dfrac{5}{6}$ to an improper fraction.

A. $\dfrac{46}{5}$

B. $\dfrac{53}{6}$

C. $\dfrac{38}{6}$

D. $\dfrac{46}{8}$

21. Add $\dfrac{5}{9} + \dfrac{5}{12}$.

A. $\dfrac{10}{21}$

B. $1\dfrac{1}{4}$

C. $\dfrac{35}{36}$

D. $\dfrac{17}{18}$

22. Subtract $\dfrac{15}{16} - \dfrac{5}{6}$.

A. 1

B. $\dfrac{5}{48}$

C. $\dfrac{1}{16}$

D. $\dfrac{3}{8}$

23. Multiply $\frac{11}{12} \times \frac{2}{33}$.

   A. $\frac{22}{45}$

   B. $\frac{2}{3}$

   C. $\frac{1}{18}$

   D. $1\frac{1}{12}$

24. Divide $\frac{9}{10} \div \frac{3}{5}$.

   A. $1\frac{1}{2}$

   B. $\frac{2}{3}$

   C. $\frac{27}{50}$

   D. $\frac{12}{15}$

25. Add $5\frac{1}{6} + 3\frac{3}{4} + 2\frac{5}{8}$.

   A. $10\frac{4}{9}$

   B. $11\frac{13}{24}$

   C. $11\frac{1}{2}$

   D. $10\frac{3}{8}$

26. Subtract $8\frac{1}{5} - 3\frac{3}{4}$.

   A. $5\frac{9}{20}$

   B. $5\frac{11}{20}$

   C. $4\frac{1}{20}$

   D. $4\frac{9}{20}$

27. Multiply $4\frac{2}{3} \times 5\frac{3}{4}$.

    A. $26\frac{5}{6}$

    B. $20\frac{1}{2}$

    C. $22\frac{1}{8}$

    D. $24\frac{5}{6}$

28. Divide $20\frac{2}{5} \div 10\frac{1}{2}$.

    A. $2\frac{4}{35}$

    B. $1\frac{33}{35}$

    C. $1\frac{19}{35}$

    D. $2\frac{1}{8}$

29. How many pieces of pipe $2\frac{1}{4}$ inches long can be cut from a piece that is $13\frac{1}{2}$ inches long?

    A. 5
    B. 7
    C. 6
    D. 8

30. Simplify $2\frac{1}{2} \div 4 \times 3 - 5$.

    A. $-4\frac{19}{24}$

    B. $\frac{6}{25}$

    C. $-3\frac{1}{8}$

    D. $\frac{5}{14}$

31. In the number 18.63278, the place value of the 7 is:

    A. tenths
    B. hundredths
    C. thousandths
    D. ten-thousandths

32. **Name the number 0.0043.**

    A. forty-three millionths
    B. forty-three thousandths
    C. forty-three ten-thousandths
    D. forty-three hundred-thousandths

33. **Round 0.23714 to the nearest thousandth.**

    A. 0.237
    B. 0.24
    C. 0.23
    D. 0.2

34. **Add 0.625 + 3.64 + 0.003.**

    A. 3.147
    B. 4.268
    C. 1.642
    D. 2.714

35. **Subtract 0.8301 − 0.0333.**

    A. 0.6824
    B. 0.7968
    C. 0.0365
    D. 0.8634

36. **Multiply 0.003 × 0.714.**

    A. 0.214200
    B. 0.021420
    C. 0.000214
    D. 0.002142

37. **Divide 2.3552 ÷ 0.512.**

    A. 46
    B. 0.46
    C. 4.6
    D. 0.046

38. **Arrange in order from smallest to largest: 0.023, 0.04, 1.2, 0.52.**

    A. 0.023, 0.04, 0.52, 1.2
    B. 1.2, 0.52, 0.04, 0.023
    C. 0.52, 0.04, 0.023, 1.2
    D. 0.04, 0.52, 0.023, 1.2

39. **Change $\dfrac{25}{33}$ to a decimal.**

    A. $0.7\overline{5}$
    B. $0.\overline{75}$

    C. 0.75

    D. 0.$\overline{758}$

40. **Change 0.32 to a fraction in lowest terms.**

    A. $\dfrac{5}{8}$

    B. $\dfrac{16}{50}$

    C. $\dfrac{8}{25}$

    D. $\dfrac{32}{100}$

41. **Multiply $0.78 \times \dfrac{3}{8}$.**

    A. $\dfrac{119}{400}$

    B. $\dfrac{14}{25}$

    C. 0.6235

    D. 0.2925

42. **If a ruler costs $0.49 and a pen costs $2.98, find the cost of 3 pens and 4 rulers.**

    A. $10.90

    B. $13.36

    C. $9.78

    D. $24.29

43. **Write 0.024 as a percent.**

    A. 2.4%

    B. 24%

    C. 0.24%

    D. 0.024%

44. **Write $\dfrac{17}{25}$ as a percent.**

    A. 6.8%

    B. 68%

    C. 0.68%

    D. 680%

45. **Write 42% as a fraction.**

    A. $\dfrac{21}{50}$

    B. $\dfrac{8}{15}$

C. $\dfrac{5}{12}$

D. $\dfrac{2}{5}$

46. **Find 32% of 138.**
    A. 431.25
    B. 4.16
    C. 43.125
    D. 44.16

47. **75% of what number is 225?**
    A. 30
    B. 16.875
    C. 300
    D. 168.75

48. **21 is what percent of 30?**
    A. $66.\overline{6}\%$
    B. 70%
    C. 7%
    D. 66%

49. **The finance rate charged on a credit card is 1.5%. Find the finance charge if the balance is $300.00.**
    A. $45
    B. $4.50
    C. $450
    D. $0.45

50. **If the sales tax on $420.00 is $23.10, find the sales tax rate.**
    A. 5%
    B. 4.5%
    C. 5.5%
    D. 6.5%

51. **Evaluate $-4y + 3xy$ when $x = -3$ and $y = 7$.**
    A. -51
    B. -21
    C. -42
    D. -91

52. **Multiply $-5(4x - 6y + 11)$.**
    A. $-20x + 30y - 55$
    B. $20x - 30y + 55$
    C. $9x - 11y + 6$
    D. $9x + 11y - 6$

53. Combine like terms: $4c + 2a - 5b + 6a - 2c + 3b$.

   A. $4a + 8b - 6c$
   B. $8a - 2b + 2c$
   C. $2a - 4b + 3c$
   D. $8a + 2b - 2c$

54. Combine like terms: $5(2x - 10) - 3(7x - 5)$.

   A. $31x - 56$
   B. $-11x - 35$
   C. $15x - 6$
   D. $20x - 30$

55. Find the Celsius temperature (°C) when the Fahrenheit temperature (°F) is 68°.
    Use $°C = \dfrac{5}{9}(°F - 32)$.

   A. $42°$
   B. $35°$
   C. $20°$
   D. $12°$

56. Solve $4x - 9 = 35$.

   A. 10
   B. 6
   C. -5
   D. 11

57. Solve $-3(2x - 6) + 16 = 22$.

   A. 6
   B. -4
   C. 2
   D. 1

58. If the sum of two times a number and 10 is 60, find the number.

   A. 5
   B. 8
   C. 15
   D. 25

59. The ratio of 15 to 6 is:

   A. $\dfrac{2}{5}$

   B. $\dfrac{5}{2}$

C. $\dfrac{2}{3}$

D. $\dfrac{3}{2}$

60. Find the value of x when $\dfrac{5}{8} = \dfrac{x}{32}$.

   A. 20
   B. 8
   C. 16
   D. 24

61. If a person uses 16 gallons of gasoline to travel 304 miles, how many gallons of gasoline will be needed to travel 361 miles?

   A. 215 gallons
   B. 19 gallons
   C. 12 gallons
   D. 27 gallons

62. Find the circumference of a circle whose radius is 24 inches. Use $\pi = 3.14$.

   A. 75.36 inches
   B. 37.68 inches
   C. 150.72 inches
   D. 113.04 inches

63. Find the perimeter of a triangle whose sides are 12 inches, 15 inches, and 18 inches.

   A. 22.5 inches
   B. 27 inches
   C. 33 inches
   D. 45 inches

64. Find the perimeter of a square whose side is 9.6 inches.

   A. 38.4 inches
   B. 92.16 inches
   C. 19.2 inches
   D. 46.08 inches

65. Find the perimeter of a rectangle whose length is 16.5 feet and whose width is 11 feet.

   A. 181.5 feet
   B. 272.25 feet
   C. 55 feet
   D. 27.5 feet

66. Find the area of a circle whose diameter is 6 yards. Use $\pi = 3.14$.

    A. 113.04 yd²
    B. 9.42 yd²
    C. 28.26 yd²
    D. 18.84 yd²

67. Find the area of a square whose side is $5\frac{3}{4}$ inches.

    A. 33.0625 in.²
    B. 23 in.²
    C. $25\frac{9}{16}$ in.²
    D. $32\frac{1}{16}$ in.²

68. Find the area of a trapezoid whose height is 22 inches and whose bases are 7 inches and 14 inches.

    A. 462 in.²
    B. 231 in.²
    C. 203 in.²
    D. 406 in.²

69. Find the volume of a sphere whose radius is 24 inches. Use $\pi = 3.14$.

    A. 57,876.48 in.³
    B. 100.48 in.³
    C. 1,356.48 in.³
    D. 32 in.³

70. Find the volume of a cylinder whose height is 16 feet and whose radius is 2 feet. Use $\pi = 3.14$.

    A. 100.48 ft³
    B. 56.52 ft³
    C. 113.04 ft³
    D. 200.96 ft³

71. Find the volume of a pyramid whose base is 12 inches by 15 inches and whose height is 8 inches.

    A. 768 in.³
    B. 720 in.³
    C. 480 in.³
    D. 1,440 in.³

72. Find the length of the hypotenuse of a right triangle if its sides are 10 yards and 24 yards.

    A. 15 yards
    B. 20 yards
    C. 18 yards
    D. 26 yards

73. How many square feet are in 1,872 square inches?

    A. 13 ft²

    B. 156 ft²

    C. 52 ft²

    D. 16 ft²

74. Change 9 yards to inches.

    A. 108 in.

    B. 324 in.

    C. 432 in.

    D. 1,296 in.

75. Change 72 feet to yards.

    A. 2 yd

    B. 8 yd

    C. 24 yd

    D. 18 yd

76. Change 9 feet 7 inches to inches.

    A. 108 in.

    B. 115 in.

    C. 96 in.

    D. 92 in.

77. Change 75 pounds to ounces.

    A. 600 oz

    B. 900 oz

    C. 1,200 oz

    D. 1,500 oz

78. Change 1,424 ounces to pounds.

    A. 71 lb

    B. 89 lb

    C. 94 lb

    D. 108 lb

79. Change 6.4 tons to ounces.

    A. 204,800 oz

    B. 102.4 oz

    C. 200,000 oz

    D. 12,800 oz

80. Change 9.3 quarts to pints.

    A. 5.64 pt

    B. 37.2 pt

C. 2.325 pt

D. 18.6 pt

81. Change 276 months to years.

   A. 3312 yr

   B. 15 yr

   C. 18.5 yr

   D. 23 yr

82. Change 4 miles to feet.

   A. 6,912 ft

   B. 10,560 ft

   C. 15,240 ft

   D. 21,120 ft

83. In which quadrant is the point (–4, –7) located?

   A. Q I

   B. Q II

   C. Q III

   D. Q IV

84. Which equation represents a horizontal line?

   A. $x = -5$

   B. $4x - 7y = 21$

   C. $y = 8$

   D. $2x + y = 6$

85. Find y when $x = -5$ for $7x + 4y = 9$.

   A. 11

   B. 7

   C. 6

   D. –7

86. Which is a solution to $2x - 3y = 9$?

   A. (4, 0)

   B. (6, 1)

   C. (–6, –1)

   D. (0, 3)

87. Find the slope of the line containing two points whose coordinates are (6, –2) and (–7, 3).

   A. $\dfrac{13}{5}$

   B. $-\dfrac{5}{13}$

C. $-\dfrac{13}{5}$

D. $\dfrac{5}{13}$

88. **Find the slope of a line whose equation is 12x + 7y = 15.**

    A. $\dfrac{7}{12}$

    B. $-\dfrac{7}{12}$

    C. $-\dfrac{12}{7}$

    D. $\dfrac{12}{7}$

89. **The slope of a vertical line is:**

    A. 0

    B. 1

    C. −1

    D. undefined

90. **Find the y intercept of the line 4x − 9y = 18.**

    A. (4.5, 0)

    B. (0, 2)

    C. (−4.5, 0)

    D. (0, −2)

91. **Add $(3x^2 + 5x - 2) + (2x^2 - 6x + 10)$.**

    A. $5x^2 - 11x + 12$

    B. $5x^2 - x - 12$

    C. $5x^2 - x + 8$

    D. $5x^2 + 11x - 8$

92. **Subtract $(12x - 3y - 11) - (15x - 6y + 9)$.**

    A. $3x - 3y - 2$

    B. $-3x + 3y - 20$

    C. $-3x - 9y + 20$

    D. $27x - 9y - 20$

93. **Multiply $8x^3 \cdot 6x \cdot (-2y^2)$.**

    A. $-96x^4y^2$

    B. $-48x^3y^2$

    C. $48xy^4$

    D. $96xy^2$

94. Find $(-5x^2y^3)^3$.
    A. $25x^8y^9$
    B. $-125x^6y^9$
    C. $25x^5y^6$
    D. $125x^6y^9$

95. Multiply $5x(2x - 9y + 5)$.
    A. $10x^2 - 45xy + 25x$
    B. $10x - 45xy + 25$
    C. $10x^2 + 45xy + 25$
    D. $10x - 45xy + 25x$

96. Multiply $(7x + 5)(3x - 2)$.
    A. $21x^2 + 29x + 10$
    B. $21x^2 - 14x - 10$
    C. $21x^2 + x - 10$
    D. $21x^2 - 10$

97. Find $(8x - 5)^2$.
    A. $64x^2 - 25$
    B. $64x^2 - 40x + 25$
    C. $64x^2 + 40x + 25$
    D. $64x^2 - 80x + 25$

98. Multiply $(5x^2 + 3x - 2)(x + 7)$.
    A. $5x^3 + 32x^2 + 23x - 14$
    B. $5x^3 - 32x^2 - 23x + 14$
    C. $5x^3 - 38x^2 - 19x + 14$
    D. $5x^3 + 38x^2 + 19x - 14$

99. Divide $64x^4y^3z^5 \div (-2x^2y^3z^4)$.
    A. $-32x^2z$
    B. $-32x^2yz$
    C. $-32x^2y^2z^2$
    D. $-32x^2yz^3$

100. Divide $21x^3y^4z^2 - 15x^4yz^3$ by $-3xy$.
    A. $-7xyz + 5xz$
    B. $7x^2y^3z^2 + 5x^3y^3z^3$
    C. $-7x^2y^3z^2 + 5x^3z^3$
    D. $7x^2yz^2 - 5x^3z^3$

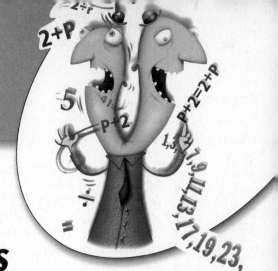

Answers to Quizzes and Final Exam

Chapter 1	Chapter 2	Chapter 3	Chapter 4
1. B	1. A	1. A	1. B
2. A	2. B	2. B	2. A
3. D	3. D	3. C	3. D
4. A	4. C	4. A	4. C
5. D	5. C	5. D	5. A
6. A	6. A	6. B	6. D
7. C	7. D	7. D	7. B
8. A	8. B	8. A	8. A
9. D	9. A	9. C	9. D
10. C	10. D	10. C	10. C
11. B	11. C	11. D	11. D
12. B	12. C	12. A	12. C
13. A	13. B	13. B	13. A
14. D	14. A	14. D	14. B
15. C	15. B	15. A	15. D
16. B	16. C	16. B	16. C
17. A	17. A	17. A	17. B
18. B	18. D	18. D	18. A
19. D	19. B	19. A	19. D
20. C	20. C	20. B	20. C

Chapter 5
1. D
2. B
3. B
4. A
5. C
6. D
7. A
8. C
9. C
10. D
11. B
12. A
13. D
14. C
15. B
16. A
17. C
18. C
19. D
20. D

Chapter 6
1. B
2. C
3. A
4. D
5. A
6. B
7. B
8. D
9. C
10. A
11. D
12. C
13. A
14. A
15. C

16. B
17. D
18. A
19. B
20. C

Chapter 7
1. B
2. D
3. A
4. C
5. C
6. A
7. D
8. A
9. D
10. B
11. C
12. C
13. A
14. D
15. B
16. C
17. D
18. A
19. C
20. B

Chapter 8
1. C
2. B
3. A
4. D
5. B
6. B
7. D
8. A
9. C

10. C
11. A
12. B
13. C
14. D
15. D
16. A
17. B
18. A
19. B
20. C

Chapter 9
1. C
2. A
3. B
4. D
5. A
6. C
7. C
8. B
9. D
10. A
11. D
12. B
13. C
14. B
15. D
16. A
17. C
18. A
19. B
20. D

Chapter 10
1. B
2. A
3. D
4. C

5. C
6. B
7. D
8. A
9. B
10. A
11. B
12. D
13. C
14. A
15. B
16. D
17. C
18. A
19. C
20. A

Chapter 11
1. C
2. A
3. D
4. B
5. C
6. C
7. B
8. A
9. D
10. D
11. C
12. B
13. C
14. A
15. C
16. D
17. B
18. A
19. A
20. C

Chapter 12
1. D
2. B
3. A
4. C
5. C
6. B
7. A
8. D
9. A
10. C
11. B
12. A
13. D
14. C
15. B
16. D
17. B
18. A
19. C
20. B

Final Exam
1. A
2. D
3. B
4. A
5. C
6. B
7. A
8. C

9. B
10. D
11. A
12. C
13. B
14. D
15. A
16. C
17. B
18. D
19. A
20. B
21. C
22. B
23. C
24. A
25. B
26. D
27. A
28. B
29. C
30. C
31. D
32. C
33. A
34. B
35. B
36. D
37. C
38. A
39. B

40. C
41. D
42. A
43. A
44. B
45. A
46. D
47. C
48. B
49. B
50. C
51. D
52. A
53. B
54. B
55. C
56. D
57. C
58. D
59. B
60. A
61. B
62. C
63. D
64. A
65. C
66. C
67. A
68. B
69. A
70. D

71. C
72. D
73. A
74. B
75. C
76. B
77. C
78. B
79. A
80. D
81. D
82. D
83. C
84. C
85. A
86. B
87. B
88. C
89. D
90. D
91. C
92. B
93. A
94. B
95. A
96. C
97. D
98. D
99. A
100. C

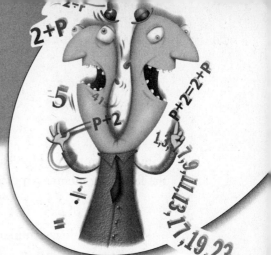

appendix

Overcoming Math Anxiety

Welcome to the program on overcoming math anxiety. This program will help you to become a successful student in mathematics. It is divided into three parts. Part I will explain the nature and causes of math anxiety. Part II will help you overcome any anxiety that you may have about mathematics. Part III will show you how to study mathematics and how to prepare for math exams. As you will see in this program, the study skills that you need to be successful in mathematics are quite different from the study skills that you use for other classes. In addition, you will learn how to use the classroom, the textbook, and yes, even the teacher, to increase your chances of success in mathematics.

Part I: The Nature and Causes of Math Anxiety

Many students suffer from what is called "math anxiety." Math anxiety is very real, and it can hinder your progress in learning mathematics. Some of the physical symptoms of math anxiety include:

- Nervousness
- Pounding heart
- Rapid breathing
- Sweating

- Nauseousness
- Upset stomach
- Tenseness

In addition to the physical symptoms, people may experience any or all of the following mental symptoms:

- A feeling of panic or fear
- Cloudy or fuzzy thinking
- Lack of concentration
- A mental block in thinking
- Feelings of helplessness, guilt, shame, inferiority, or stupidity

If you have any of these physical or mental symptoms when you are in a mathematics classroom, during a mathematics test, or when you are doing your mathematics homework, then you suffer some degree of math anxiety.

Naturally, a little fear or uneasiness accompanies all of us when we take exams, but if these symptoms are severe enough to keep you from doing your best, it is time to do something about your math anxiety.

In order to decrease your anxiety, it is helpful to determine why you have math anxiety. After all, no student, as far as we know, is born with math anxiety. Let's look at some possible causes of math anxiety, and while you are reading them, think about some reasons that might have caused you to develop math anxiety.

Reason 1: Poor Math Teachers

Throughout elementary school, middle school, and even in high school, you have had many mathematics teachers. Some were good, and some were not. Sometimes teachers are forced to teach math when they are not trained for it or when they dislike the subject themselves, or even when they do not understand it. Naturally, in these situations, teachers cannot do a good job. Think about your teachers. Were they good or not?

Reason 2: Traumatic Experiences

Many students can recall an incident in their education involving mathematics where they had a traumatic experience. Were you ever called to the blackboard to do a math problem in front of the whole classroom and were unable to do it?

Did your teacher belittle you for your inability in mathematics? Were you ever punished for not doing your mathematics homework?

Sometimes teachers, parents, or tutors can make you feel stupid when you ask questions. One such traumatic experience in a person's life can cause that person to fear or hate mathematics for the rest of his or her life.

One student recalled that when he was studying fractions in third grade, the teacher picked him up by his ankles and turned him upside down to illustrate the concept of inverting fractions!

Can you recall a traumatic event that happened to you that was related to mathematics?

Reason 3: Prolonged Absence from School

Students who miss a lot of school usually fall behind in mathematics. Many times, they fall so far behind that they cannot catch up and end up failing the course. This is because mathematics is a cumulative subject. What you learn today you will use tomorrow. If you fail to learn a topic because you were absent, when the time comes to use the material to learn something new, you will be unable to do so. It's like trying to build a second story on a building with a weak or inadequate first story. It cannot be done. For example, if you look at long division, it involves multiplication and subtraction. So if you can't multiply or subtract, you will not be able to do long division. All mathematics is like this!

Think about a time when you were absent from school. Were you lost in math class when you came back?

Reason 4: Poor Self-Image

Occasionally, students with a poor self-image have difficulty with mathematics. This poor self-image could have been acquired when students were told something like this:

"Men can do math better than women."

"Math requires logic, and you're not very logical."

"You should be able to do math in your head."

"Boy, that's a stupid question."

These comments can be made by parents, teachers, spouses, and even friends. They can make a person feel very inadequate when doing mathematics.

I have heard that it takes 10 positive comments to overcome one negative comment.

Have you ever been made to feel stupid by someone's negative comment?

Reason 5: Emphasis on the Correct Answer

Mathematics involves solving problems, and solving problems involves getting the correct answers. When doing mathematics, it is very easy to make a simple mistake. This, in turn, will lead to the wrong answer. Even when you use a calculator, it is very easy to press the wrong key and get an incorrect answer.

Furthermore, when students are under pressure in a test situation, they become nervous and tend to make more mistakes than they would on their homework exercises. When this occurs, students may feel that it is impossible to learn mathematics. This can cause a great deal of anxiety.

Have you ever failed a test, even when you knew how to do the problems but made simple mistakes?

Reason 6: Placement in the Wrong Course

In college, it is easy to sign up for the wrong mathematics course—that is, a course in which you do not have enough math skills to succeed. Therefore, it is necessary to be sure that you are properly prepared for the course that you are going to take.

Most mathematics courses, except the introductory ones, have **prerequisites**. For example, in order to succeed in algebra, you need to have an understanding of arithmetic. Success in trigonometry requires knowledge of algebra and geometry.

Before signing up for a mathematics course, check the college catalog to see if the course has a prerequisite. If so, make sure that you have successfully completed it. **Do not skip courses.** If you do, you will probably fail the course. It would be like trying to take French III without taking French I and French II.

If you are enrolled in a mathematics course now, have you completed the prerequisite courses? If you are planning to take a mathematics course, did you check the college catalog to see if you have completed the prerequisite courses?

Reason 7: The Nature of Mathematics

Mathematics is unlike any other course in that it requires more skills than just memorization. Mathematics requires you to use analytical reasoning skills, problem-solving skills, and critical thinking skills. It is also abstract in nature since it uses symbols. In other words, you have to do much more than just memorize a bunch of rules and formulas to be successful in mathematics.

Many students view mathematics with the attitude of, "Tell me what I have to memorize in order to pass the test." If this is the way you think about mathematics, you are on the road to failure. It is time to change your way of thinking about mathematics. From now on, realize that you have to do far more than just memorize. How do you view mathematics?

Of course, there are many other factors that can cause math anxiety and lead to poor performance in mathematics. Each person has his or her own reasons for math anxiety. Perhaps you would like to add your own personal reasons for your math anxiety.

Part II: Overcoming Math Anxiety

Now that we have explained the nature of math anxiety and some of the possible causes of it, let us move to the next part of this program: How You Can Overcome Math Anxiety. The plan for overcoming math anxiety consists of two parts: **First,** you need to develop a **positive attitude** toward mathematics. **Second,** you need to learn some **calming** and **relaxation skills** to use in the classroom, when you are doing your homework, and especially when you are taking a math test.

Developing a Positive Attitude

To be successful in mathematics, you need to develop a positive attitude toward the subject. Now you may ask, "How can I develop a positive attitude for something I hate?"

Here's the answer.

First, it is important to **believe in yourself**. That is, believe that you have the ability to succeed in mathematics. Dr. Norman Vincent Peale wrote a book entitled *You Can If You Think You Can*. In it he states, "Remember that self-trust is the first secret of success. So trust yourself." This means that in order to succeed, you must believe absolutely that you can.

"Well," you say, "how can I start believing in myself?"

This is possible by using affirmations and visualizations.

An **affirmation** is a positive statement that we make about ourselves. Remember the children's story entitled *The Little Engine That Could*? In the story, the little engine keeps saying, "I think I can. I think I can" when it was trying to pull the cars up the mountain. Although this is a children's story, it illustrates the use of affirmations. Many athletes use affirmations to enhance their performance. You can, too. Here are a few affirmations you can use for mathematics:

"I believe I can succeed in mathematics."

"Each day I am learning more and more."

"I will succeed in this course."

"There is nothing that can stop me from learning mathematics."

In addition, you may like to write some of your own. I think you can see what I mean. Here are some suggestions for writing affirmations:

1. State affirmations in the positive rather than the negative. For example, don't say, "I will not fail this course." Instead, affirm, "I will pass this course."

2. Keep affirmations as short as possible.

3. When possible, use your name in the affirmation. Say, "I (your name) will pass this course."

4. Be sure to write down your affirmations and read them every morning when you awake and every evening before going to sleep.

Affirmations can also be used before entering math class, before doing your homework, and before the exam. You can use affirmations to relax when your symptoms of math anxiety start to occur. This will be explained later.

Another technique that you can use to develop a positive attitude and help you in believing in yourself is to use **visualizations.** When you visualize, you create a mental picture of yourself succeeding in your endeavor. Whenever you visualize, try to use as many of your senses as you can—sight, hearing, feeling, smelling, and touch. For example, you may want to visualize yourself getting an A on the next math test. Paint a picture of a scene like this:

Visualize yourself sitting in the classroom while your instructor is passing back the exams. As he or she gives you your exam, feel the paper in your hand and see the big red A on top of it. Your friend next to you says, "Way to go!" I don't know how to include your sense of smell in this picture, but maybe you can think of a way.

The best time to visualize is after relaxing yourself and when it is quiet. Of course, you can use visualization any time when it is convenient.

A fourth way to develop a positive attitude is to give yourself a **pep talk** every once in a while. Pep talks are especially helpful when you are trying to learn difficult topics, when you are ready to give up, or when you have a mental block.

How we talk to ourselves is very important in what we believe about ourselves. Shad Helmstetter wrote an entire book on what people say to themselves. It is entitled *The Self-Talk Solution*. In it he states, "Because 75% or more of our early programming was of the negative kind, we automatically followed suit with self-programming of the same negative kind." As you can see, negative self-talk will lead to a negative attitude.

Conversely, positive self-talk will help you develop a positive self-image. This will help you not only in mathematics but also in all aspects of your life.

Another suggestion to help you to be positive about mathematics is to be **realistic.** You should not expect to have perfect test papers every time you take a test. Everybody makes mistakes. If you made a careless mistake, forget it. If you made an error on a process, make sure that you learn the correct procedure before the next test. Also, don't expect to make an A on every exam; sometimes a C is the best that you can accomplish. In mathematics, some topics are more difficult than others, so if you get a C or even a lower grade on one test, resolve to study harder for the next test so that you can bring up your average.

Finally, it is important to develop **enthusiasm for learning.** Learning is to your mind what exercise is to your body. The more you learn, the more intelligent you will be. Don't look at learning as something to dread. When you learn something new and different, you are improving yourself and exercising your mind. Learning can be fun and enjoyable, but it requires effort on your part.

Developing a positive attitude about mathematics alone will not enable you to pass the course. Achieving success requires hard work by studying, being persistent, and being patient with yourself. Remember the old saying, "Rome wasn't built in a day."

Learning Calming and Relaxation Skills

The second part of this program is designed to show you some techniques for stress management. These techniques will help you to relax and overcome some of the physical and mental symptoms (nervousness, upset stomach, etc.) of math anxiety. The first technique is called **deep breathing.** Here you sit comfortably in a chair with your back straight, feet on the floor, and hands on your knees. Take a deep, slow breath in through your nose and draw it down into your stomach. Feel your stomach expand. Then exhale through your mouth. Take several deep breaths. Note: If you become dizzy, stop immediately.

Another technique that can be used to calm yourself is to use a **relaxation word.** Sit comfortably, clear your mind, and breathe normally. Concentrate on your breathing, and each time you exhale, say a word such as "relax," "peace," "one," or "calm." Select a word that is pleasing and calming to you.

You can also use a technique called **thought stopping**. Any time you start thinking anxious thoughts about mathematics, say to yourself, "STOP," and then try to think of something else. A related technique is called **thought switching**. Here you make up your mind to switch your thoughts to something pleasant instead of thinking anxious thoughts about mathematics.

In the preceding section, I explained the techniques of **visualization** and **affirmations**. You can also use visualizations and affirmations to help you reduce your anxiety. Whenever you become anxious about mathematics, visualize yourself as being calm and collected. If you cannot do this, then visualize a quiet, peaceful scene such as a beautiful, calm lake in the mountains. You can also repeat affirmations such as "I am calm."

The last technique that you can use to relax is called **grounding**. In order to ground yourself, sit comfortably in a chair with your back straight, feet on the floor, and hands on your legs or on the side of the chair. Take several deep breaths and tell yourself to relax. Next, think of each part of your body being grounded. In other words, say to yourself, "I feel my feet grounded to the floor." Then think of your feet touching the floor until you can feel them touching. Then proceed to your ankles, legs, etc., until you reach the top of your head. You can take a deep breath between grounding the various parts of the body, and you can even tell yourself to relax.

In order to learn these techniques, you must practice them over and over until they begin to work.

However, for some people, none of these techniques will work. What then? If you cannot stop your anxious thoughts, then there is only one thing left to do. It is what I call the **brute force technique**. That is to realize that you must attend class, do the homework, study, and take the exams. So force yourself mentally and physically to do what is necessary to succeed. After brute forcing several times, you will see that it will take less effort each time, and soon your anxiety will lessen.

In summary, confront your anxiety head on. Develop a positive attitude and use stress reduction techniques when needed.

Part III: Study and Test-Taking Techniques

Having an understanding of math anxiety and being able to reduce stress are not enough to be successful in mathematics. You need to learn the basic skills of how to study mathematics. These skills include how to use the classroom, how to use the textbook, how to do your homework, how to review for exams, and how to take an exam.

The Classroom

In order to learn mathematics, it is absolutely necessary to **attend class**. You should never miss class unless extreme circumstances require it. If you know ahead of time that you will be absent, ask your instructor for the assignment in advance, then read the book, and try to do the homework before the next class. If you have an emergency or are ill and miss class, call your instructor or a friend in the class to get the assignment. Again, read the material and try the problems before the next class, if possible. Be sure to tell your instructor why you were absent.

Finally, if you are going to be absent for an extended period of time, let your instructor know why and get the assignments. If you cannot call him or her, have a friend or parent do it for you.

Come to each class prepared. This means having all necessary materials including homework, notebook, textbook, calculators, pencil, computer disk, and any other supplies you may need.

Always select a seat in front of the classroom and near the center. This assures that you will be able to see the board and hear the instructor.

Pay attention at all times and take good notes. Write down anything your instructor writes on the board. If necessary, bring a tape recorder to class and record the sessions. Be sure to ask your instructor for permission first, though.

You can also ask you instructor to repeat what he or she has said or to slow down if he or she is going too fast. But remember, don't become a pest. You must be reasonable.

Be sure to ask intelligent questions when you don't understand something. Now I know some of you are thinking, "How can I ask intelligent questions when I don't understand?" or "I don't want to sound stupid."

If you really understand the previous material and you are paying attention, then you can ask intelligent questions.

You must also remember that many times the instructor will leave out steps in solving problems. This is not to make it difficult for you, but you should be able to fill them in. You must be alert, active, and pay complete attention to what's going on in class.

Another important aspect you should realize is that in a class with 10 or more students, you cannot expect private tutoring. You cannot expect to understand everything that he or she teaches you. But what you must do is copy down everything you can. Later, when you get home, apply the information presented in the chapters of this book.

Be alert, active, and knowledgeable in class.

The Textbook

The textbook is an important tool in learning, and you must know how to use it.

It is important to **study** your book. Note that I did not say "read" your book, but I purposely said **study.**

How do you study a mathematics book? It is different from studying a psychology book. First, look at the chapter title. It tells you what you will be studying in the chapter. For example, if Chapter 5 is entitled "Solving Equations," this means that you will be doing something (solving) to what are called "equations." Next read the chapter's introduction. It will tell you what topics are contained in the chapter.

Now look at the section headings. Let's say Section 5.3 is entitled "Solving Equations by Using the Multiplication Principle." This tells you that you will be using multiplication to solve a certain type of equation.

Take a pencil and paper and underline in the book all definitions, symbols, and rules. Also, write them down in your notebook.

Now actually work out each problem that is worked out in the textbook. Do not just copy the problems, but actually try to solve them, following the author's solutions. Fill in any steps the author may have left out. If you do not understand why the author did something, write a note in the book and ask your instructor or a friend to explain it to you. Also notice how each problem is different from the previous one and what techniques are needed to find the answer. After you have finished this, write the same problems on a separate sheet of paper and try to solve them without looking in your book. Check the results against the author's solutions.

Don't be discouraged if you cannot understand something the first time you read it. Read the section at least three times. Also, look at your classroom notes. You may find that your instructor has explained the material better than your book. If you still cannot understand the material, do not say, "This book is bad. I can't learn it." What you can do is go to the library and get another book and look up the topic in the table of contents or appendix. Study this author's approach and try to do the problems again. There is no excuse. If the book is bad, get another one.

Remember that I didn't ever say that learning mathematics was easy. It is not, but it can be done if you put forth the effort!

After you have studied your notes and read the material in the textbook, try to do the homework exercises.

Homework

Probably the single most important factor that determines success in mathematics is doing the homework. There's an old saying that "Mathematics is not a spectator sport." What this means is that in order to learn mathematics, you must do the homework. As stated previously, it is like learning to play an instrument. If you went to the music class but never practiced, you could never learn how to play your instrument. Also, you must practice regularly or you will forget or be unable to play your instrument. Likewise, with mathematics, you must do the homework every day it is assigned. Here are my suggestions for doing your homework:

- First and most important: **DO YOUR HOMEWORK AS SOON AS POSSIBLE AFTER CLASS.** The reason is that the material will still be fresh in your mind.

- Make a habit of studying your mathematics regularly, say, three times a week, five times a week, etc.

- Get your book, notes, and all previous homework problems, calculator, pencil, paper (everything you will need) before you start.

- Do not dally around. Get started at once and do not let yourself be interrupted after you start.

- Concentrate on mathematics only!
- Write the assignment on the top of your paper.
- Read the directions for the problems carefully.
- Copy each problem on your homework paper and number it.
- Do not use scratch paper. Show all of your work on your homework paper.
- Write neatly and large enough. Don't do sloppy work.
- Check your answers with the ones in the back of the book. If no answers are provided, check your work itself.
- See if your answers sound reasonable.
- Write out any questions you have about the homework problems and ask your instructor or another student when you can.
- Draw pictures when possible. This is especially important in courses such as geometry and trigonometry.
- If you have made a mistake, try to locate it. Do not depend on the teacher to find all your mistakes. Make sure that you have copied the problem correctly.
- Don't give up. Doing a problem wrong is better than not doing it at all. (Note: Don't spend an exorbitant amount of time on any one problem though.)
- Use any of the special study aids, such as summaries, lists of formulas, and symbols that you have made.
- Don't skip steps.
- If you cannot get the correct answer to a problem, don't stop. Try the next problem.

Review

In order to learn mathematics, it is necessary to review before the tests. It is very important to realize one fact. You cannot cram in mathematics. You cannot let your studying go until the night before the exam. If you do, forget it. I have had students who have told me that they spent 3 hours studying before the exam, and then failed it. If those were the only 3 hours they spent studying, there is no way they could learn the material.

Some teachers provide written reviews. Make sure you do them. If this is the case, you can use the review as a practice test. If not, you can make up your own review. Many books have practice tests at the beginning and the end of chapters. If so, you can use these exams as reviews. Some books have extra problems at the end of each chapter. By doing these problems, you have another way to review.

Finally, if there is no review in the book, you can make up your own review by selecting one or two problems from each section in the chapter. Use these problems to make up your own practice test. Be sure not to select the first or second problem in each unit because most mathematics books are arranged so that the easy problems are first.

When you review, it is important to memorize symbols, rules, procedures, definitions, and formulas. In order to memorize, it is best to make a set of cards as shown here:

Front	**Back**
Commutative Property of Addition	For any real numbers, a and b, a + b = b + a

On the front of the card, write the name of the property, and on the back, write the property. Then when you are studying, run the cards through the front side and then on the other side. This way you can learn both the property and also the name of the property.

Finally, you must be aware that a review session is not a study session. If you have been doing your work all along, then your review should be short.

Test-Taking Tips

There are three types of exams that mathematics instructors give. They are closed book exams, open book exams, and take home exams.

First, let's talk about closed book exams given in the classroom. Make sure that you show up 5 to 10 minutes before the class begins. Bring all necessary materials such as a pencil with an eraser, your calculator, paper, and anything else that you may need. Look over and study materials before class. After you receive your exam paper, look over the entire test before getting started. Read the directions carefully. Do all of the problems that you know how to do first, and then go back and try the others. Don't spend too much time on any one problem. Write down any formula that you may need. Show all necessary steps if a problem requires it.

If there is any time left after you finish the exam, go back and check it for mistakes.

If you don't understand something, ask your instructor. Remember you cannot expect your instructor to tell you whether or not your answer is correct or how to do the problem.

In open book exams, you should remember that the book is your tool. Do not plan to study the book while you are taking the test. Study it before class. Make sure you know where all the tables, formulas, and rules are in the book. If you are permitted, have the formulas, rules, and summaries written down. Also, have sample problems and their solutions written down.

Don't be elated when your instructor gives you a take home exam. These exams are usually the hardest. You may have to go to the library and get other books on the subject to help you do the problems, or you may have to ask other students for help. The important thing is to get the correct solution. Show all of your work.

When your instructor returns the exam, make a note of the types of problems you missed and go back and review them when you get home.

See your instructor for anything that you are not sure of.

A Final Note

This program explained the nature and causes of math anxiety and showed some ways to reduce or eliminate the symptoms of math anxiety. The last part of the program explained study, review, and test-taking skills needed for success in mathematics. But there is still one problem left to discuss. What happens if you are in over your head? What happens if the material you are studying is still too difficult for you?

First, try getting a math tutor. Many schools provide tutors through the learning center or math lab. Usually the tutoring is free. If your school does not provide free tutors, then seek out one on your own and pay him or her to help you learn. Also, many textbooks have online teaching sessions. You can use these as tutoring sessions.

There are other things you can do to help yourself if tutoring doesn't work. You can drop the course you are in and sign up for a lower-level course next semester. You can also audit the course. That is, you can take the course but do not receive a grade. This way you can learn as much as you can but without the pressure. Of course, you will have to enroll in the course again next semester for a grade.

Many colleges and some high schools offer non-credit brush-up courses such as algebra review or arithmetic review. If you need to learn the basic skills, check your school for one of these courses. Usually no tests or grades are given in these courses.

I hope that you have found the suggestions in this program helpful, and I wish you success in your mathematical endeavor.

Index

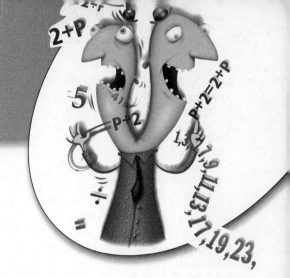